SAVING THE BREEDS

A HISTORY OF THE RARE BREEDS SURVIVAL TRUST

EXMOOR PONIES

Other books by Lawrence Alderson include:

A Chance to Survive (Jolly 1989 and 1978)

Other books by Valerie Porter include:

Pigs: A Handbook to the Breeds of the World (Helm Information 1993)
Cattle: A Handbook to the Breeds of the World (Helm 1992)
Practical Rare Breeds (Pelham 1987)
The Southdown Sheep (Southdown Sheep Society 1991)

SAVING THE BREEDS

A HISTORY OF
THE RARE BREEDS SURVIVAL TRUST

By
Lawrence Alderson
and
Valerie Porter

Line drawings by Sue Franklin

Pica
PRESS

© 1994 Lawrence Alderson and Valerie Porter

Helm Information Ltd

The Banks, Mountfield,

Near Robertsbridge, East Sussex TN32 5JY

ISBN 1-873403-30-5

The Pica Press is an imprint of Helm Information Ltd

Designed and Typeset by Fluke Art,
Bexhill on Sea, East Sussex

Printed and Bound by
Hartnolls Limited, Bodmin, Cornwall

CONTENTS

PLATES

1. Rare Breeds Task Conference, October 1971. Peter Jewell presents his paper. The meeting was chaired by Sir Dudley Forwood Bt, flanked by Ann Wheatley-Hubbard and Christopher Dadd

2. Chairman of the Trust, Michael Rosenberg, and fund-raising officer, Marshall Watson, discuss the Trust's exhibit at the Royal Show in 1982 with Minister of State for Agriculture, Peter Walker (later Lord Walker)

3. Denis Vernon presents the trophy for the champion Lincoln Longwool ram, Billingborough Caesar, to Harold Nobes BEM, shepherd for breed chairman, Robert Watts, at the Volvo-sponsored National Show and Sale in 1988

4. Interbreed pig judge, Geoffrey Cloke, and Viki Mills with one of her Berkshire entries at the National Show and Sale in 1986. Viki took the Interbreed Championship with a Middle White gilt

5. Michael Rosenberg receives the CBE from Minister of State for Agriculture, Michael Jopling, in 1986. Also in the picture, left to right, are Denis Vernon, Sir Richard Cooper Bt, Lord Benstead (Minister of State MAFF), Alastair Dymond, Joe Henson, John Hawtin, John Hearth (Chief Executive of RASE) and Geoffrey Cloke

6. HRH Prince Philip being welcomed on the Trust stand at the Royal Show in 1989 by Her Grace The Duchess of Devonshire. Staff members Val Nicholson and Kay Burgess join in the welcome

7. A reception at the Trust's demonstration unit near entrance 3 at the National Agricultural Centre. From right, J.D.F Green, Christopher Dadd (foreground), Shell U.K representative, John Lightbown (at rear), Michael Rosenberg, Mary Dadd, Richard Ferens (Hon. Director of RASE), Ann Wheatley-Hubbard

8. Paul Hooper of Masterbreeders displays a tube of boar semen from the Trust's flask at a workshop in 1987. Geoffrey and Jonathan Cloke are on the left of the picture

9. The owners of the winning pair of Longhorn cattle are presented with the Burke Trophy by HM The Queen at the Royal Show in 1981

10. Anne Petch, South-West regional winner of the NFU Mutual Marketing Award Scheme in 1984, receives her award from Minister of State for Agriculture, John MacGregor

11. A group of dignitaries at the 1991 AGM at Beamish. Left to right: front row, Denis Vernon (Treasurer), Mrs Teresa Wickham (Director of Corporate Affairs at Safeways), Lord Elliott of Morpeth (retiring President); back row, Sir Richard Cooper, Bt. (Chairman) and Sir Derek Barber (later Lord Barber of Tewkesbury) (President-elect)

12. Anne Petch being greeted by the Prime Minister at 10 Downing Street in 1987

13. Pat Cassidy, Editor of *The Ark* since 1983, at her desk in the Trust's new offices at the National Agricultural Centre

14. Council member Anne Petch (left) and Technical Consultant Lawrence Alderson (behind) at the 1991 National Show and Sale with Silvija Davidson, winner of the 1990 Masterchef competition with a rare breeds recipe, and David Natt

15. Members of Council at a meeting in January 1981 at 35 Belgrave Square line up in front of a historic painting of the RASE. Left to right: seated, Robin Otter, Lawrence Alderson (Technical Consultant), Michael Rosenberg (Chairman), Denis Vernon, Christopher Dadd; standing, John Hawtin, John Cator, Christopher Marler, Richard Cooper (later Sir Richard Cooper, Bt), Peter Jewell, David Steane and Denys Stubbs

16. Sale of rare breeds at Bite Farm in 1983. Nancy Briggs in the ring

17. The AGM in 1988 was held in the delightful setting of Temple Newsam Estate, one of the most important centres in the Trust's Approved Centre scheme. Breeds were paraded for the large audience in a custom-built amphitheatre

18. Crowds throng the Trust's marquee at the Great Yorkshire Show in 1982

19. HRH The Duke of Gloucester inspects the Trust's major demonstration in the Farming Heritage feature at the Devon County Show in 1975. He is hosted by Lawrence Alderson (left) and Michael Rosenberg

20. Michael Heseltine lends his voice to the Trust's fund-raising effort at the Celebration of Food & Farming event in 1989 in Hyde Park. Alongside are Sir Richard Cooper, Bt and Her Grace The Duchess of Devonshire

21. Chairman of the Trust, Dudley Reeves, greeting HRH Princess Alexandra at the Trust stand at the second Hyde Park event in 1992. Merchandising officer, Rosalind Ragg, is in the background

22. A rare opportunity to relax at the National Show & Sale; Show Director, John Hawtin, and Technical Consultant, Lawrence Alderson, in 1985

23. Wyllie Turnbull judging a British Lop class at the National Show & Sale in 1979. Geoff Collings is parading his entry, Geoffrey Cloke is the ring steward, and John Backhouse is in the centre of the group of spectators

24. Left to right: Sir Dudley Forwood, Bt, Ann Wheatley-Hubbard (Chairman of the Trust), John Hawtin (in-coming Director) and Michael Rosenberg (retiring Director) at the National Show & Sale in 1982

25. An Approved Centre inspection workshop held in 1992 at Norwood Farm, owned by member of Council, Cate Mack. Left to right: Geoffrey Cloke (vice-president), Brian Brooks (chairman of Farm Parks sub-committee), Dudley Reeves (chairman of the Trust), Lawrence Alderson (Executive Director), Peter King (Field Officer), Jeremy Roberts (member of Council) and (seated) Andrew Sheppy (member of Council)

26. A family group at Wimpole Home Farm; managed by Bernard Hartshorn (centre) and Shirley Hartshorn (right), it maintains important units of many rare breeds including Longhorn cattle and Bagot goats

27. The Trust's Patron, HRH The Prince of Wales, opened the International Conference at Warwick University in 1989. Here he is welcomed by Denis Vernon, Chairman of the Trust. Also in the picture are John Hodges (left) and Sir Richard Cooper, Bt

ILLUSTRATIONS

WHITE PARK BULL

PREFACE

There is inevitably a sense of satisfaction when an ambition is realised, and a feeling of fulfilment from the part played in that achievement. Such sentiments are heightened even further when the influence engendered is significant and ongoing. As a member of the Working Party that pioneered a crusade and founded the Rare Breeds Survival Trust, I am very proud and happy, more than twenty years later, to admit my susceptibility to such emotions. I expect that they will be clearly evident in the pages that follow but, more important, I hope that this history pays tribute adequately to the immense and prime role of the small group of people who initially developed the concept and launched the movement for the conservation of endangered breeds of livestock in Britain, as well as to the increasing body of devotees who follow in their path. It has been my privilege to know and respect as friends most of the initial group of founders; people of great character - Sir Dudley Forwood Bt, Bill Stanley, the late Christopher Dadd, Ann Wheatley-Hubbard, Idwal Rowlands and others. We all owe them a very special debt of gratitude.

The research involved in writing this history aroused a flood of old memories, many of which had lain dormant for years, or which had been suppressed by more highly publicised stories. It was essential that they were committed to print without delay. Memories fade too quickly and recollections become distorted. The current surge of international activity is heralding a new era which would obscure even more the events of the past if they are not recorded properly now. No longer is the Trust a fringe organisation; it now commands a large and global audience and exerts a powerful influence in many fora.

The growth and success of the Trust shows no sign of abating. Its programmes continue to expand and its reputation as the leader in its field is undiminished. The founders now have been succeeded by the current generation of leaders who have secured their own place in history. I hope that the publication of past achievements now will inspire the next generation who will take the Trust to new heights. Their success will serve to enhance further the vision and achievements of their predecessors. My life has been enriched immeasurably by my association with the Trust, in terms both of experiences and of friendships. I am sure that it has been, and will be, the same for many others.

I am grateful to Christopher Helm for his interest in undertaking the publication of the History and for his faith in the subject, and to Valerie Porter who has confirmed again her reputation, not only as an entertaining writer, but also as an indefatigable pursuer of essential information. I have valued very much her support and constructive contributions as co-author. Finally, I am grateful to all the members and supporters of the Trust, too numerous to permit individual mention, but without whom this history could never have been written.

Lawrence Alderson   April 1994

CHAPTER ONE

THE EARLY DAYS

SHROPSHIRE SHEEP

PORTLAND SHEEP

CHAPTER 1:
THE EARLY DAYS

GESTATION

The history of the Rare Breeds Survival Trust can, indirectly, be traced back to 1825 when Sir Stamford Raffles proposed the formation of a Zoological Society with the object of 'introducing and domesticating new Breeds or Varieties of Animals ... likely to be useful in Common Life' with the intention of finding new 'varieties, breeds, and races of animals for the purpose of domestication or for stocking our farm-yards, woods, pleasure-grounds, and wastes ...'

Raffles became the first president of the Zoological Society of London when it was formed the following year. It soon opened its zoological gardens on rented land at Regents Park. A hundred years later the place was bursting with more than 5,000 animals on display and the ZSL established a 'holiday camp' for surplus, sick or breeding stock at Hall Farm, Whipsnade, on the Bedfordshire downs. Domesticated breeds of farm livestock were included at both sites.

It can fairly be claimed that the germ of the Rare Breeds Survival Trust was born on the several estates of wealthy and often eccentric landowners but was nurtured on those downs at Whipsnade. This zoological connection was one of the major strands that, quite by chance, would come together when the moment was ripe to be neatly interwoven in the form of the early rare breeds movement during the 1950s and 1960s.

The Whipsnade Gene Bank

In 1955 Solly Zuckerman became Honorary Secretary of the ZSL and immediately began to take practical steps to realise a dream born of a nightmare. During the Second World War he had witnessed the devastation of European livestock, slaughtered in large numbers for food and also indiscriminately by the concentrated acts of warfare itself. In Britain, large acreages of pasture were ploughed up in the race to grow food crops, inevitably at the expense of livestock. Zuckerman realised that there was a danger of many traditional breeds becoming so low in numbers as to risk extinction, and he also appreciated that breeds, in the same way as the species of wild animals that were his society's direct concern, needed positive conservation if they were to survive. He considered that Whipsnade could provide a haven for some of the traditional farm breeds: there were already a few cattle and sheep and there was plenty of space for more, especially on the self-contained Home Farm within the Park.

Zuckerman was encouraged in this by one of his ZSL Council members in particular — D.V. Davies, editor of *Grays Anatomy* and, as it happens, the uncle of Philip Ryder-Davies, who later became veterinary surgeon to London Zoo and developed (independently) a deep interest in rare breeds of farm livestock. Ryder-Davies is still well known in rare breed circles today and, in his own quiet way, has been a crucial figure in the movement.

There is an apocryphal tale that Zuckerman's inspiration for the Whipsnade collection of rare breeds was the Scotch or Scots Dumpy hen. Battery-cage systems were becoming the fashion and in these a chicken's legs were more of a hindrance than a help. The Dumpy was known for its very short legs but it was already a very rare breed and indeed thought to be extinct, until a few were found in east Africa. Zuckerman pointed out that, however unlikely it might seem, many other minority breeds might have the potential to meet agricultural needs of the future which at present could not even be imagined. Who, in their right mind, would have designed a stumpy-legged chicken in the days when free-ranging was the birds' typical lifestyle?

In reality, it was cattle that encouraged Zuckerman to take an initiative in 1957. After an abortive attempt two years earlier to house some Charolais cattle offered as a gift, he contacted the Hannah Dairy Research Unit in Ayr and offered Whipsnade (which had quarantine facilities) as a 'holding camp' for the breeds which the unit was importing for its climatological experiments. Clearly he saw a role for Whipsnade in accommodating interesting domestic species as well as wild ones. In 1959 he took this concept a step further: on 24 July he wrote to the Rt Hon John Hare, Minister of Agriculture, in the following terms:

> I wonder whether I can interest you in a not-too-scatterbrained idea which I put to J.H. Ritchie, your Chief Veterinary Officer, about a year ago, and to which he was very sympathetic.
>
> As Secretary (Hon.) of the Zoological Society of London, I have asked myself whether we could use the resources of Whipsnade Park (near Dunstable) to maintain breeding units of types of sheep and cattle which are now disappearing. For example, the Norfolk Big Horned Sheep and the Cotswold Sheep have all but disappeared. If they become extinct, their particular gene characteristics also disappear, and hundreds of generations might pass before they crop up again. My idea is, in effect, to establish at Whipsnade a 'library' of rare genes. I spoke to Sanders about this the other day, and he, too, was taken by the idea, and asked why I hadn't spoken to you about it. Hence this letter.
>
> In addition to trying to build up breeding units of disappearing ruminants, I am also hoping that we shall be able to do the same for some types of African cattle which are now being eliminated because they are less efficient than others. This is a matter which I have also discussed, and shall discuss again, with Ritchie, in view of the f. & m. limitation. Since we are licensed for quarantine purposes, I have also suggested to the Hannah people who are working on environmental adaptations of different ungulates for the Colonial Office, that we could act as a holding centre for tropical cattle which they are bound to want to investigate as their research progresses.
>
> If we can proceed with this idea, I hope I shall have your blessing when I get in touch with the N.A.A.S., who are probably the best people to say where in this country one can find the remnants of disappearing breeds of sheep and cattle.

This letter marks the first serious consideration of preserving what were rapidly becoming endangered breeds of livestock. Would that the Rare Breeds Survival Trust had been formed at that very moment! If it had been, it could have saved pig breeds like the Lincolnshire Curly Coat and the Dorset Gold Tip ....

By the end of 1959 the ZSL Council had been informed that Zuckerman was 'examining the question of disappearing breeds of domestic animals'. Coincidentally, the Council had created a new Breeding Policy Committee (BPC), a group of specialists in the biology of reproduction. Among those in attendance was E.H. Tong, director of Whipsnade, and the committee was chaired by Professor Alan S. Parkes. Its secretary would soon be Dr Idwal W. Rowlands, the first director of the new Wellcome Institute of Comparative Physiology set up at Regents Park shortly after the committee's formation; and an early member of the BPC would be Peter Jewell, the Institute's first appointment, who soon took up a research fellowship to study the breeding biology of Soay sheep on the island of St Kilda.

The BPC first met on 6 October 1959 and in considering its terms of reference Zuckerman pointed out that it was important not only to breed wild animals but to establish a 'gene library' of rare and disappearing breeds of domestic animals. The committee noted that there was plenty of space at Whipsnade for domestic breeds: if necessary, a hundred acres could be available. They intended to investigate what should be done experimentally to increase breeding success in general for their collection of both wild and domestic animals and would consider using all modern techniques to remedy any infertility.

In October 1960 a joint meeting of the BPC and the ZSL's Collections Policy Committee (chaired by the Earl of Cranbrook) confirmed the formation of the Gene Library at Whipsnade, not as part of the public zoological exhibition in the Park but tucked away at the Home Farm on a new area of grassland in large subdivided paddocks with a suitable regimen of rotational grazing and fallowing. The collection would not normally be open to the public except on certain days, advertised in advance (perhaps six times a year), but it would be available to scientific workers and others with a professional interest in the animals, by arrangement only. However, the fringes of the territory could be accessible to the public so that a selection of animals could be seen close to the boundary. It was hoped that, as a later development, there might be a 'genetic exhibition to show the variety produced in domestication of animals'. It was stated very clearly that the future organisation of Whipsnade would be in three parts: a zoological Park containing the wild animals, a Gene Library containing breeding stocks of disappearing domestic breeds, and a Home Farm (for growing fodder and other crops) which would diminish in size as the Gene Library grew.

The Gene Library was formally established at Whipsnade in February 1961 and its first livestock breeds included the remnants of the old Norfolk Horn sheep and a group of Manx Loghtan sheep. The Manx Loghtan flock at Whipsnade had come from their home island to Regents Park in 1955, two years after the Manx Museum had begun to maintain its own native flock.

The name of the new collection was quickly changed to the Gene Bank, a deliberate evocation of 'blood bank' as it would be 'a depository of genes from which they could be recovered when required'. At this stage it was decided that, because of the large number of domestic breeds worldwide, it should contain only British breeds in the first instance. In order to put the proposal into effect, the first step should be an investigation into existing breeds, their present location and availability, a task in which it was decided to enlist the help of whatever breed societies might exist for them. Tong, as Director of Whipsnade Park, undertook to obtain further information on rare and disappearing British mammalian breeds, while Alan Parkes would do the same for poultry as he already had a long-standing interest in the subject and plenty of information to hand. Initially it was decided to include cattle, sheep and poultry; pigs might be considered during a second phase, and perhaps horses and ponies in a third phase. Their enthusiasm was such that they even began to wonder about other species apart from farm animals — for example, rabbits, guinea pigs, dogs and pigeons. They set about finding and acquiring suitable stock and by February 1962 Tong was able to table a detailed report of the Gene Bank's poultry, sheep and cattle.

In the summer of 1963 Zuckerman asked the BPC for a definition of the scope of the Gene Bank and, if possible, its cost to the Society — which was perhaps something of a warning bell. The committee decided to set up a Gene Bank subcommittee (GBSC) whose terms of reference were to make recommendations on: Gene Bank policy (with particular reference to the breeds of British cattle, sheep and poultry to be maintained); the area of Whipsnade Park to be set aside for them; the additional staff, if any, required; and the estimated annual cost to the society. The members of the GBSC were to be Parkes as chairman, Tong as Director of Whipsnade Park, Dr P.A. Jewell and Dr I.W. Rowlands. Jewell, with practical farming experience and an agricultural degree as well as having read physiology at Cambridge, had been interested in primitive island sheep since the 1940s; Rowlands had come to the Wellcome Institute from the Institute of Comparative Physiology at Babraham. The four were ideally suited to looking after the interests of rare breeds and they also invited Dr Alan Robertson of the Institute of Animal Genetics, Edinburgh, to attend their first meeting.

Early in the 1960s there was a change of staff at Whipsnade. Tong, a land agent when he joined in 1947, had been using a local veterinarian on a part-time basis to help with any problems among the stock. It was decided that it would be sensible to have a full-time veterinarian as deputy director and an advertisement to this effect was seen by Victor Manton, a lively and good-humoured man with an agricultural background, who had been evacuated as a child to Dunstable and who used to spend his spare time at Whipsnade for pleasure before he qualified from veterinary college in the early 1950s. He took up the new post in March 1962, and was promoted in 1968 (after Tong's retirement) to become Curator of Whipsnade, a post he held until 1989.

The BPC's decision not to include pigs in the Gene Bank was somewhat ironic in that it had let slip a golden opportunity to acquire the last of the Lincolnshire Curly Coat pigs. In the early 1960s Philip Ryder-Davies had stayed with Lord Cranbrook (who had founded the Mammal Society, as well as being chairman of the ZSL's Collections Policy Committee) and they discussed vanishing breeds of farm animals. The Earl mentioned that the pig breeds in particular were being neglected and Ryder-Davies decided to undertake a 'little superficial research'. He reported his findings to Cranbrook in a letter in November 1963, stating that he had found six breeds that seemed to be in a sorry state. He could locate no breeders at all of the Dorset Gold Tip, the Yorkshire Blue and White or the National Long White Lop (he had not yet penetrated the fastnesses of Cornwall!); there were only four breeders of the Cumberland, three of the Oxford Sandy and Black and one of the Curly Coat. 'As I was going to a wedding in Lincolnshire on Saturday,' he wrote of the latter breeder, 'I arranged to go and spend the day with him — a Mr Crowder, of Market Stainton, near Lincoln. Incidentally, it was fascinating to find that he also bred Percheron horses and Lincoln Longwool sheep, which are also falling in popularity.' And then came his suggestion: would the ZSL be interested in acquiring a few of the Curly Coats for Whipsnade? Apart from two or three recently sold by Crowder, his were certainly the last of the breed, though unfortunately he had been inbreeding them for some time. Incidentally, Ryder-Davies was only a medical student at the time, with exams looming.

Spreading the net

In April 1964 the journal *Nature* (vol. 202) published an article by Rowlands which explained in detail the history and aims of the Gene Bank and invited greater interest in the project from the scientific community in general. Writing to the same journal in November 1971, the then Lord Zuckerman, as Secretary of the ZSL, set out the Gene Bank story up to that period thus:

> On a decision taken by the Society's council, small flocks and herds of rare native domesticated breeds were collected and established at Whipsnade in 1961 with the object of preserving them, not merely as historic remnants, but as material for study by geneticists, physiologists and other scientists. An offer to put the animals at the disposal of *bona fide* investigators and to collaborate in their research projects was made in your journal in 1964. A useful study of the blood groups of sheep was carried out by Tucker; otherwise the response to our offer was disappointing, and we failed either to stimulate interest or to obtain financial support for the appointment of scientists to our own staff to study these animals.

They began to wonder why they had a gene bank at all. They were also beginning to despair at their lack of success generally in breeding or improving the fertility of some of the Gene Bank animals, especially the Norfolk Horns. They would try a more direct way of generating interest in December 1966, which happened to be the year when Deryk Frazer was elected to the committee.

Frazer was a naturalist with a lifelong passion for British fauna in general and an interest in rare breeds. In the 1930s as a school boy he had visited the newly opened Whipsnade collection, where his attention was caught by pairs of Chillingham and White Park cattle. Trained as a medical doctor, he switched to physiology and later became wholly involved with wildlife conservation. He was Director of Nature Conservancy in England; he also bought a small farm and raised sheep and pedigree cattle.

When Frazer joined the BPC, he took a particular interest in the rare sheep breeds he discovered at Whipsnade. He felt that they were in the wrong environment and would fare better if they were managed elsewhere as farm livestock rather than zoo animals. He was to be an important link between the government in the guise of the Nature Conservancy, and the Mammal Society of which he was a founder member along with Cranbrook, and the future Rare Breeds Survival Trust, though he deliberately avoided being a member in its first few years because he felt he could be of more help as an outsider.

In 1966 he put forward a detailed and imaginative paper about the gene bank, submitting it to Idwal Rowlands and Peter Jewell in the first instance. It asked very basic questions such as: 'What is the purpose of holding these stocks?' His own reply was that there was a strong case for keeping a reserve of genetic material which may well prove of major value to agriculturalists in the future, as well as of zoological interest. How should they be maintained and by whom: zoos, agriculturalists or educationalists? Frazer thought the RASE was the sort of body which should be expected to give financial support. He made out a particularly strong case for the educational value of the breeds. They would demonstrate to agricultural students and farmers, as well as the general public, their relationships to modern breeds. He considered that the existing facilities and finances of the Gene Bank were inadequate and wondered what the best policies for future breeding and expanding the existing stocks would be, and whether there were other breeds in need of conservation — or indeed other species: why were there no pigs and no horses in the Gene Bank? Should a second country estate be acquired for Gene Bank animals? Should there be a full-time shepherd and stockman, working to a full-time research worker who might spend at least part of the time in the countryside actively searching for other relict stocks? Where should the Gene Bank turn for grants and other sources of income?

Tracking them down

In the meantime, Philip Ryder-Davies had been touring the country in search of livestock breeds. It began with an interest in Jacob sheep, which he investigated with his customary thoroughness with the intention of finding the origin of every known flock and analysing it to see where the breed had come from in the first place. In July 1965 he had published an article on the subject in *The Field*. He contacted every livestock officer in every county during his private survey, and found they were all delighted that someone was taking an interest in the local breeds. He travelled widely to follow up the leads he was given, browsing through show catalogues, talking to people, taking photographs of the last known Oxford Sandy and Black purebred pigs (and some Tamworth/Berkshire crosses masquerading as OSBs). He saw the last Lincoln Curly Coats, those fat, woolly pigs which had been the smallholder's stand-by for years; he took photos of them on Crowder's farm and two or three years later he was contacted by the breed society's last secretary, who handed over the minute books for safe keeping; Ryder Davies eventually gave them to the Museum of English Rural Life in Reading. He photographed an unknown herd of Blue Albion cattle in

Wales (a breed which suffered disastrously during a foot-and-mouth outbreak in 1967) and saw the only flock of Shropshire sheep; he pinned down just six stallions of the Irish Draught horse in its homeland (there were none in the UK); he saw the original Vaynol cattle in their namesake setting. There were very few rare breeds that he did not actually see in the flesh.

Ryder-Davies was not, as it happens, the only one to realise that in order to preserve rare breeds it was necessary to know that they existed and where and how many, and how fast they were dwindling. Peter Jewell had the same notion but at first it met with little enthusiasm, especially from Tong at Whipsnade who dismissed the idea, saying that he could perfectly well rely on the Ministry's agricultural advisory officers and their regional colleagues for such information.

Looking back over quarter of a century, it is easy to say now that one of the big mistakes was to hide the rare breeds away on the Home Farm at Whipsnade, instead of having them on public display (bar a few tame pets in the children's corner and the ranging flocks on the downs). If the animals are not seen, they are forgotten. And if only a national survey of breeds had been undertaken by an organisation with the resources to do so during the early 1960s, several might have been saved which were then at a critical point.

In the meantime, another development was taking place in the heart of the Midlands which would in due course productively entangle itself with the fading Gene Bank project. It formed one of the four major strands that eventually gave birth to the RBST.

Stoneleigh

The Royal Agricultural Society of England (RASE) had always held its annual Royal Show at different venues, on a rotation through each of several districts. This merry-go-round continued until well after the second world war, by which time the 'Royal' had become such a big event that there were considerable problems in finding adequate facilities. Towards the end of the 1940s, one or two people began to think about the possibility of permanent show sites but the idea was not taken seriously until the latter half of the 1950s. When the Norwich Show produced a deficit of £20,000, it was at last decided to search for a permanent, central site.

In 1960 a subcommittee, under the chairmanship of Lt Col Sir Walter Burrell, was approached by Lord Leigh suggesting that his own Home Farm at Stoneleigh, between Coventry and Leamington Spa, might be suitable. In due course a lease was agreed, with an option for an extension of the lease once a couple of shows had been held on the site in 1963 and 1964.

It was to become much more than a site for the annual Royal Show. It would be a national agricultural centre (a new principle agreed in 1961), a place where progressive ideas and techniques would be given every opportunity to flourish and be disseminated.

Stoneleigh became a reality in 1962. The society advertised for a 'Technical Director and Secretary' and the successful applicant was a crop husbandry officer on the East Anglian staff of the National Agricultural Advisory Service (NAAS, later ADAS): Christopher Dadd, a man who can be regarded as one of the founders of the RBST. His first task, however, was to prepare for the two consecutive Royals at Stoneleigh. If they were successful, there was every hope that the site would indeed become a permanent one.

By the summer of 1965 Stoneleigh had become a permanent show site. There was a great deal to do to get the ambitious new venture of a National Agricultural Centre underway and the NAC was not officially opened until September 1967. Christopher Dadd's title would be changed to Director, NAC, on a full-time basis in 1969; meanwhile, and thereafter, one of his main responsibilities was to strengthen links with other organisations with the help of the NAC's deputy secretary, W.R.B. (Ray) Carter, seconded from the MAFF in 1966.

Drawing the threads together

The decade of the 1960s was one of change, as those who lived through it can testify. Attitudes were being altered in many aspects of life, and that included the world of the zoological collections, though hesitantly at first. For a long time zoos had existed simply to collect and exhibit the curious — the exotic — but gradually people were realising that the earth's resources of all kinds were far from infinite and that it was time to take action to reduce the wasteful exploitation that had accompanied rapid economic growth.

There was a dawning realisation that zoos had a vital role in the conservation of endangered species. They could become involved in careful breeding programmes to increase their stock and also, ultimately, perhaps return animals to the wild to repopulate their depleted natural habitats. Sir Peter Chalmers Mitchell, secretary of the ZSL in first few years of the century, had been far ahead of his time with his vision of wild animals roaming almost freely in wide open spaces at a country branch of Regents Park — the original concept for Whipsnade.

In November 1966 the United Nations' FAO (Food and Agricultural Organisation) held a Study Group in Rome about domestic animal genetic resources; it would hold another in 1968. At that stage the group deemed that domestic animals should be 'evaluated for their performance with respect to present or short-term future requirements' — in contrast to the future RBST's view that the breeds should be conserved for their own sake, as irreplaceable genetic material, regardless of current economic demands, in that the future is unknown and the breeds should be kept as insurance against possible needs.

The atmosphere was right for the seeds of the idea of conserving endangered domesticated breeds and it so happened that the events of the 1960s, at Whipsnade and Stoneleigh in particular, and also in the academic world and in the rapid development of an increasing leisure market and tourism, laid the ground for the positive combination of all these interests, propelled to some extent by the negative pressures of lack of resources at Whipsnade. The time had come and people with a mutual (albeit sometimes latent) interest in rare breeds were poised to come together.

The initiative came from the zoological strand. At Whipsnade, the continued lack of breeding success with the tiny Norfolk Horn flock and one or two other sheep breeds had become demoralising and in any case the domesticants began to seem an anachronism there. The ZSL had taken a strategic look at the whole idea of Whipsnade and felt that the rare breeds simply did not fit the place's image. More practically, they had begun to appreciate that zoos were not best equipped to manage farm livestock, nor was the land at Whipsnade particularly suitable for them. The ZSL still strongly supported the theory of a gene bank for domestic breeds but could not continue to go it alone in practice. It wanted to involve others, both in active research and in help with funding for its maintenance. The 1964 *Nature* article had not succeeded in generating that involvement.

One of several romantic tales about those good old days comes from the late Christopher Dadd: that the eventual formation of the RBST was precipitated by white rhinos. It was a series of coincidences, of people who happened to know people. Dadd had been at Stoneleigh since New Year's Day, 1963, and now that it had been confirmed as a truly permanent showground he had left his beloved mill home at Grantchester and built a new house on site. Sir Francis Pemberton would retire as Honorary Show Director after the 1967 Royal Show and the role would be taken over by Sir Dudley Forwood, Bt, who became a close friend of Dadd's. It so happened that Sir Dudley, as well as being a member of the RASE, was on the Council of the ZSL.

'Dudley came to me one day,' Dadd recalled, 'and said, "Christopher, can you help the ZSL? We have a problem. We have fields full of rare breeds in the Gene Bank at Whipsnade. And we are expecting some white rhinos from Kenya — they are literally boarding the ship at this moment and we must have an empty field for them on their arrival. The rare breeds will have to go; we might even have to convert them into lion fodder. You couldn't possibly accommodate them here at Stoneleigh, could you?"'

Idwal Rowlands, today and at the time, expressed indignation at the suggestion of lion fodder: the society would never, he said, have killed off perfectly healthy animals just because they had run out of space. Zuckerman himself, in his 1971 letter in *Nature*, said:

> What I should like to correct is the statement [made in 'Keeping Fossils Alive', a report on a conference on rare breed survival held on 15th October, 1971] that the Society dispersed the gene bank it had established on its own initiative because 'space was needed'. The animals were sent to institutions where it was hoped they would attract more interest than they had done at Whipsnade and where they would consequently be of greater value to agricultural science.

Memory does play false tricks. In reality the major white rhino importation from Natal was not even suggested until December, 1969, by which time Whipsnade's poultry and sheep had already found new homes — not because of lack of space so much as because agricultural contexts were deemed more appropriate for them. When the Natal Game Parks Administration indicated that it would wish to dispose of about a hundred white rhino by the summer of 1970 it was perfectly timed, from Whipsnade's point of view, for by then the Park was making a loss and suffering from sharp falls in attendance. There was even a threat of complete closure for a while. It was decided that the white rhino project could be an extremely important development for Whipsnade, as well as for the white rhino.

In essence, the ZSL felt that the gene bank was more appropriately the concern of the agricultural and scientific community than that of a zoological park. Whipsnade, it was felt, had played its part: it had created the bank and thereby 'rescued' a few of the breeds. Now it was someone else's turn to extend the concept.

In December 1966 Zuckerman therefore wrote to Christopher Dadd at the RASE (enclosing a copy of Rowlands' *Nature* article) proposing a joint meeting between the ZSL's Gene Bank subcommittee and other bodies such as ARC, NAAS, MAAF and Reading University as well as the RASE. This would be a turning point for rare breeds; and indeed Rowlands's article had already sown the seed by acting as a focus for people who had always had a vague or consuming interest in traditional breeds. A movement was about to be born.

BIRTH

In his letter of 1 December 1966 to Christopher Dadd, Zuckerman wrote:

> The Society is anxious that our Committee which looks after the Gene Bank project at Whipsnade should meet representatives of various scientific bodies having an interest in the preservation of rare breeds of farm animals.
>
> The purpose of such a meeting would be to create a wider interest and to find out how far other organisations could assist in this work. We should appreciate, for instance, having opinions on the general usefulness of the project, as it now stands, to agricultural science and whether it

should be extended to include other species of farm animals (horses, pigs), or by increasing the number of breeds of the animals we already have at Whipsnade.

The Society is unable to do more than maintain the project at its present level, so that if any extension of it appears to be justifiable we shall have to know what financial help we could look to from other bodies.

We are greatly hampered by the lack of information on the changing status and whereabouts of these so-called rare breeds in this country, and one solution to this problem that we should like to discuss is the possible employment of an agricultural graduate for one or more years to survey and report on the situation. I am certain it would be very rewarding to have this information made readily available to all concerned.

We would be delighted if you and another member of your staff having expert knowledge on this subject could attend the meeting, here in the Zoo, in the early part of next year. If so, I will pass on your reply to Rowlands, who is the Secretary of the Breeding Policy Committee, to arrange a date convenient to all concerned.

The enclosed reprint provides something of the aims of the project, but should you require any more recent information about it, I am sure Rowlands would be pleased to help.

Spreading the load

Idwal Rowlands, a modest and delightful man, was the driving force behind the organisation of this first joint exploratory meeting (and, quietly, behind many other factors which would lead to the development of the RBST) which was held on 16 February 1967, in the ZSL Council Room at Regents Park. The meeting was designed to bring other organisations more actively into the scheme and to consider broadly whether the gene bank animals at Whipsnade were usefully preserved, whether the breeds were disappearing anyway, and, if it was agreed that they should be kept, who should keep them. Frazer's paper had had an effect on everybody's thinking within the ZSL: they had heeded his pressure for the care of the rare breed flocks to be passed to more appropriate organisations with a better understanding of livestock management.

It was an interesting group that gathered at that first joint meeting. It was chaired by the well known ABRO geneticist, Dr Alan Robertson, in view of his special qualifications in the genetics of breeding. Rowlands and Tong were there, along with C.G.C. Rawlins (Director of Zoos) and Dr R.M. Sadleir (Rowlands's colleague at the Wellcome Institute). Christopher Dadd represented the NAC, and Deryk Frazer the Nature Conservancy. Robert Trow-Smith, at the time assistant editor of *Farmer & Stockbreeder* and the author of two major books on the history of livestock breeds, was present and also Bill Longrigg from the Ministry of Agriculture, a man who had first become interested in British White cattle some twenty years before. There was also Reading University's newest professor, John C. Bowman, with considerable knowledge of crossbreeding programmes and the consequences of inbreeding in animals; he had worked with dairy cattle and then in the poultry industry for eight years with a company which had brought together a collection of about a hundred poultry strains from around the country as a gene pool for future use, a concept not unlike that of the Whipsnade gene bank.

The attendance of Dadd, Trow-Smith, Longrigg and Bowman stressed the gradual and logical swing towards agriculture, while Rowlands and others provided the link with the long-standing zoological interests. The net result of that first meeting would be the dispersal of Whipsnade's sheep flocks to the NAC and to Reading University, switching the onus firmly away from zoology and on to agriculture. Zuckerman, in his 1971 *Nature* letter, continued the story thus:

> After exchanges with other scientific bodies, and following much deliberation by its scientific committees, the council of the Society was advised that greater use might be made of these valuable stocks if they were transferred to agricultural institutions where they would attract the interest of students, and where better facilities for rearing the young would be available. Accordingly, in 1968, flocks of three breeds of sheep were transferred to Professor Bowman at the School of Agriculture, University of Reading. Arrangements were made with Mr Christopher Dadd, Director of the National Agricultural Centre, for the transfer of the remaining four breeds of sheep to Stoneleigh Park, Kenilworth, where they have now attracted considerable public interest. In 1970, the NAC was also presented with surplus animals from the Zoological Society's herd of Chartley Cattle at Whipsnade.

A long memorandum drafted by Bowman covered the general agreements made at the joint meeting in February 1967. Entitled 'The case for maintaining "control" populations of almost extinct breeds of domestic livestock', it set out the basic principles which would underlie the rare breed movement and are still relevant — indeed vital — today, and it gave detailed proposals as to how the stated aims of saving the disappearing breeds might be achieved. Its first proposal was for the appointment of a co-ordinator who was a qualified biologist with particular interests in genetics, evolution and comparative biological studies and who had particular abilities in the field of public relations. The co-ordinator would develop a 'preservation' project based on the existing Whipsnade gene bank and would travel widely about the country to raise support and to supervise populations of the breeds which would be established at various centres where there might be suitable land — mention was made of universities, agricultural societies with permanent showgrounds, the Nature Conservancy and 'other organisations concerned with conservation who might have spare land'. The paper ended with a list of the rare breeds then held at Whipsnade:

		Adult stock	
		m	f
Cattle	Longhorn	3	12
	* Chartley	1	12
Sheep	Cotswold	1	11
	Norfolk-Horned	4	7
	Lincoln Longwool	1	21
	Manx Loghtan	2	2
	Woodlands White-faced	1	9
	Portland	1	8
	Soay	12	10
Poultry	Silver Spangled Hamburgs	3	8
	Lakenvelders	4	9
	Sumatra Game Fowl	1	3
	Indian Game Fowl		5
	Old English Game Fowl	1	14
	Buff Cochins	1	9
	Redcaps	3	8
	Silver Dorkings	4	11

[*Note that Bowman still uses the name Chartley, a consistent error at Whipsnade: the animals are more properly White Park as a breed, not Chartley as a particular type within the breed.]

Bowman's long memo was presented to the Agricultural Research Council (ARC) as an appropriate body, with a suggestion that ARC should support the future management of the Gene Bank sheep and cattle or, if this was not possible, should find others able to give financial support.

That crucial joint meeting in February 1967 had agreed in principle upon the need to preserve the sheep and cattle of the Whipsnade gene bank. The June meeting of the old GBSC agreed that its poultry could be disbanded to private individuals and most of them were (or had already been) dispersed to the Duchess of Devonshire, with the remainder to the Universities Federation for Animal Welfare.

A year after the joint meeting, the Earl of Cranbrook, as chairman of the Federation of British Zoos (he was a man of many parts), took a positive step by writing on 19 February 1968 to Sir Peter Greenwell, of Woodbridge, Suffolk, with a personal plea for the preservation of 'vanishing' breeds and suggesting that the NAC might be a good home for them. The net result was a meeting between the RASE and Whipsnade on 13 May 1968 to investigate the possibility of the RASE taking over some of the Whipsnade sheep. An added spur to the RASE's reaction was that the ARC had said it was not in a position to take responsibility for the cost of developing the Gene Bank and the investigational work involved. Greenwell and Canon Peter Buckler (Director of Demonstrations) represented the RASE at the May meeting; the ZSL was represented by Rawlins, Rowlands as Secretary of the BPC, and Tong as Director of Whipsnade. It was a productive meeting, as Buckler's notes after the event suggest:

MEETING AT WHIPSNADE ZOO - MONDAY, 13TH MAY, 1968

This meeting followed a suggestion by Sir Peter Greenwell, that we should investigate the possibility of taking over the sheep now kept at Whipsnade. This suggestion was put to him by the Earl of Cranbrook in a letter dated 19th February, 1968.

These sheep were the subject of a meeting held at Belgrave Square on the 16th February, 1967, when the future of a Gene bank was discussed.

As a result of that meeting, it was suggested that the Agricultural Research Council might be responsible for the cost of developing the Gene bank and the investigational work involved. It seems that since this date the A.R.C. has indicated that it is not in a position to do this. If the proposals had gone ahead they would have covered cattle and sheep, but not poultry.

At Whipsnade we met Mr C.G.C. Rawlins, Director of Zoos, Dr I.W. Rowlands, Secretary of The Breeding Policy Committee, and Mr E.H. Tong, Director of Whipsnade Park.

The present position is that they have the following breeds of sheep:-

Cotswold Sheep	14
Norfolk-Horned	15
Lincoln Longwool	29
Manx Loghtan	7
Woodlands White-faced	16
Portland	15
Soay	75

There are eight cross-bred Norfolk x Suffolk.

It is felt that the Norfolk-Horned and Manx Loghtan breeds are in danger of extinction, the others might be preserved.

The recommendations that follow are based on thoughts that one has had in mind for a number of years, and on arguments that could be developed at length. For the sake of brevity, I am listing the recommendations.

1) The Royal Agricultural Society of England is the right body to take over these sheep, and preserve the breeds if possible.
2) We should ask the Sheep Working Party to take this development

under its wing and to plan the housing and grazing of the breeds.

3) When this has been done and an estimate of cost prepared, then we should:-

    a) Approach the Sponsors of the Sheep Centre and ask them if they are prepared to include this development under their wing.

    b) Agree with the Zoo the transfer of the breeds.

I would suggest that, at this stage, we might consider adopting the first six, but ask them to retain the Soay sheep for the time being at least, as these are wild and need very much better fencing than we are likely to have available at the start. Furthermore, there is a nucleus of this breed on St Kilda and it is unlikely to die out.

Perhaps eventually we could have some of these sheep in a pen at the Show, or as part of a farm Zoo.

Housing and Grazing

The land at Whipsnade is liable to be very wet and they have no adequate housing available. Indeed, old sheep have to be kept in boxes used for transporting animals to the Zoo.

What is needed is wintering accommodation in any dutch barn. Some of the cattle shedding or any of the other similar buildings on the Showground would be ideal for this purpose. If the breeds could be housed in the winter it would be easier for people to look at them and we would be likely to get very much better breeding results. This should be possible at Stoneleigh with very little cost.

For summer grazing some of the breeds could be run together, but preferably we would need paddocks in varying sizes. In the long-term, I would like to think that these breeds could link up with the Demonstration Units. By fencing in the grassland between the Units we should be able to graze the sheep in separate groups and it would give a link and prevent the present 'Margate in the winter' appearance of the place. Until this can be done there are paddocks that can be used, the ones that are used for the cross-bred sheep demonstration at the Show etc.

At the time of the Show the breeds could be on display in small paddocks retained for this purpose. These need not be all at one place, there are odd areas of grass which, for one reason or another, are not taken up by stands and it would be rather nice to have little units of sheep spread about the Showground to relieve the monotony of lines of stands and machinery.

Alternatively, they could be moved and run together on a field away from the Showground.

    1) They would be cared for by the Shepherd.

    2) We would need to create a small sub-committee, including Professor Bowman of Reading, Dr I.W. Rowlands, Mr W. Longrigg and others.

These would be concerned with the use of the breeds as a Gene bank, the Working Party would be responsible for the day-to-day running and management of the flock.

I am quite certain that this is worth doing. Due to circumstances which cannot be helped, conditions at Whipsnade are not satisfactory and some of the breeds may die out through environmental conditions alone. At Stoneleigh they would have a much better chance of survival.

It is suggested that this would be the beginning of a bigger involvement by the Royal Agricultural Society of England in the preservation of the lesser known breeds. Other Counties and organisations have established museums for old equipment, farm tools etc. No-one is in a better position than the Society to preserve these breeds, and even if today, and indeed, in the future, they have no great economic value, nevertheless, it must be worthwhile preserving nucleuses of them.

As far as the sheep are concerned, I gathered that the gentlemen at Whipsnade were less concerned about in-breeding than they were about the inference of environment and other factors such as low numbers in a flock on reproduction. For instance, if we run other sheep with the breeds that are only in small numbers there is a chance that the breeding cycle may be better. Furthermore, we must expect breeding through the year, and although it is not defensible scientifically, it seems that where only small numbers are available to breed in captivity there is a tendency to produce males. It might be that better housing at Stoneleigh, wider grazing and the addition of the use of other breeds to make up larger groups with a ram, might all help towards increased numbers.

The Society could also act as a clearing house for information on lesser known breeds of sheep and other farm stock. An article in the 'Review' with a form to be completed, could give us information and the addresses of folk with animals of breeds we have heard about, and possibly some that will be new to us.

The Royal Agricultural Society of England is in touch with knowledgeable farmers and others all over Gt. Britain, and could follow up this article with visits to the herds and flocks that seem to have animals of interest. It is possible that as a result of this, we might get to know of small groups of animals of some of the breeds of interest which have escaped the notice of the people at Whipsnade. Having collected this information together, the Society might form a section, or even another Society for the preservation of breeds, and encourage those farmers who have got animals of interest, to preserve them, and, indeed, others might be persuaded to take on a nucleus of these breeds, if they know that this is part of a national exercise and not just a private endeavour to keep going a breed which is quickly dying.

There are many interesting ideas and important themes in Buckler's report. In the meantime, John Bowman was still interested in moving some of the Gene Bank sheep to Reading, in spite of the ARC's failure to help. In the end there was a compromise: on 2nd August 10 Cotswolds, 12 Portlands and 14 Hirta Soays were transferred to Reading from Whipsnade, leaving the RASE with the choice of the rest. On 12 August, Rowlands and Tong from the ZSL met Buckler and Dadd to finalise the details of stock movements and also agreed that, once the RASE had taken over the animals, it was entirely up to them what should happen to the stock. Further, they agreed that attempts should be made to establish the whereabouts of other rare breeds of sheep and also of cattle and pigs, so that 'the loss of potentially valuable genes should not occur'.

On 18 September, 13 Whitefaced Woodlands, 22 Lincoln Longwools, 14 Norfolk Horns and 7 Manx Loghtans went over to Stoneleigh, where they apparently caused considerable interest — so much so that the RASE considered expanding the project and began to express an interest in Whipsnade's White Park cattle. At that stage the Whipsnade herds included 18 'Chartley' and 27 Longhorns.

Deryk Frazer was pleased at the outcome: the flocks were handed over shortly after he had completed his stint on the BPC. However, it was another couple of years before he discovered, accidentally, that a condition of the handover of sheep to the NAC was that they should 'keep in touch with Dr Frazer'. No one from either ZSL or RASE had informed him of that, or approached him, but, as he says, by then there was really nothing for him to do other than to pay a visit to Stoneleigh and observe the flocks happily grazing on good pasture.

In October 1968 the RASE released a press notice about the Gene Bank:

FARM ZOO AND GENE BANK

The first animals to form the nucleus of a 'farm zoo' and gene bank, so that the potentialities of rare breeds which might otherwise become extinct may be preserved for all time, have arrived at the National Agricultural Centre, Kenilworth.

First arrivals are sheep representing the Norfolk Horned, Lincoln Longwool, Manx Loghtan and Woodlands Whitefaced breeds transferred to the Centre from Whipsnade.

This is the result of an agreement between the RASE and the Zoological Society. Details were arranged at a meeting in August between Dr I.W. Rowlands, secretary of the Whipsnade Breeding Policy Committee, Mr E.H. Tong, Director of Whipsnade, Mr C.V. Dadd, RASE Secretary and Technical Director, and the Rev. Peter Buckler, Director of Demonstrations, RASE. Technical supervision of the gene bank will ultimately be under the direction of a consultative panel upon which a number of distinguished geneticists and zoologists are being asked to serve.

Sheep are only the beginning of a larger concept. Endeavours will be made to locate any other animals of rare breeds of cattle and pigs, as well as sheep which may join the zoo and gene bank.

There is no problem of accommodation at the Centre. Four paddocks for ewes and two for rams will be provided during the tupping season and there is ample winter housing and grazing paddocks.

The movement of the livestock from Whipsnade attracted a flood of enquiries to the NAC about endangered breeds. People were beginning to realise that the breeds' disappearance would be a loss of genetic variation, even if they did not quite express it in those terms. The movement was well and truly underway. All that remained of the gene bank at Whipsnade itself by the end of 1968 would be the Longhorn and 'Chartley' cattle. The onus on maintaining the disappearing breeds had shifted firmly from the zoological sector to the agricultural.

KERRY BULL

The Working Party

The idea of a technical consultative panel (mentioned in the press release) had been the subject of discussion between Rowlands and Dadd that summer, and during October Dadd's deputy, Ray Carter, convened a meeting at the NAC to consider it further. It was suggested that the panel would 'provide guidance on the management for scientific purposes of the existing flock' and should also consider the 'concentration of further rare breeds of domesticated livestock with the same object'. The meeting looked at other possible terms of reference for such a panel. This was the germ of a new Working Party which would look not only at the management of the NAC's new sheep gene bank but also more widely at the future of rare breeds of livestock.

The new group, under the cumbersome title of the Rare Breeds of Domestic Livestock Gene Bank Working Party, held its first meeting on 19 March 1969. Those present included some familiar names: the chair was taken by W.F. (Bill) Stanley, who was chairman of the RASE's Demonstration committee (which was setting up and running the demonstration units at the NAC); the minutes were taken by Dadd's deputy, Ray Carter; John Bowman from Reading and Bill Longrigg from the MAFF were there. Apologies for absence were received from Buckler, Dadd and Rowlands and it is perhaps a matter of regret that the latter two, who had been the major driving forces in the movement so far, could not be present. Rowlands himself said as much in a letter to Carter dated 1 April:

> It was unlucky that the three people (Dadd, Buckler and myself) who were cognisant of the events leading to the transfer of the animals to NAC and Reading were not present so that a brief historical review could have been recorded. It was most unfortunate that I developed 'flu a few days beforehand and was in no fit state to attend the meeting.

The meeting had been held at the RASE's London offices in Belgrave Square. The agenda, circulated by Carter, stated first its possible terms of reference, for discussion, as (a) to provide guidance on the management, for scientific purposes, of the existing rare breeds of sheep at NAC 'and even beyond'; (b) to consider collecting and preserving further rare breeds of domesticated livestock with the same objective; (c) any other relevant aspects. The agenda also included a look at the present position of rare breeds of cattle, sheep, pigs and poultry, and a closer one at the sheep at NAC and Reading University. The final item was: 'To consider how best to set up an organisation to preserve those breeds likely to possess desirable characteristics.' This is the first formal move to create a separate organisation concerned with rare breeds.

The main points to emerge from the meeting were the reasons for conserving animals: (a) the conservation of desirable features which formed the heritage of this country and therefore had historical value; (b) the use of specific genes within such breeds to produce desirable characteristics in modern breeding programmes. Ideas would change a little when the Trust was finally formed. They recommended the following terms of reference for the Working Party: '(a) To collect and preserve rare breeds of domesticated livestock for conservation and scientific purposes; (b) To provide detailed and general guidance, as appropriate, on the maintenance of such breeds and in particular to those in charge of such animals being kept for the specific purpose to stimulate interest and research.' Then they tried to define 'rare', in a way that would not suit the Trust today: '(a) In danger of extinction; (b) in danger of losing closed identity; (c) when the number of animals in an active breeding state are reduced to five males and twenty females.'

The Working Party's chairman, Bill Stanley, remembers now that the RASE could at that time do no more than act as a catalyst and he recalls that Christopher Dadd had pulled together 'all the parties who would be interested in preserving the breeds' to serve on the Working Party. Of the six men invited by Carter to that first meeting, the most faithful attenders at subsequent meetings would be Dadd, Rowlands and Stanley.

All sorts of little developments emerged during the summer of 1969, including lengthy correspondence about a breed that would remain controversial for at least two more decades: the Oxford Sandy and Black pig. Rare breed owners began to respond to the challenge; for example Oscar Colburn of Crickley Barrow, Gloucestershire, well known for his new Colbred sheep and later an interesting speaker on the lack of strategy within the agricultural industry as a whole, offered to donate some Cotswold ewes to the NAC gene bank — it was Colburn who had done most to keep the breed going from the original Garne flock. In the event, Colburn generously presented the NAC with six tupped ewes, saying, 'We would make no condition over these animals, they would become entirely the property of the RASE to manage and dispose of as they like.'

The second meeting of the Working Party, at the NAC, was on 24 September 1969 and by then its name had been shortened a little to Rare Breeds of Domesticated Livestock Working Party. As well as chairman Stanley and the familiar faces of Dadd, Carter, Rowlands and Bowman, the members now also included John Farmer (in his final year as farms manager at NAC) and Dr Russell Jones of the ZSL, an Australian and a new research fellow under Rowlands at the Wellcome Institute — his special knowledge concerned wool growth and ram semen, and he soon became involved in the Norfolk Horn breeding programme.

They discussed the MAFF's offer to report on the position of rare breeds nationwide, Gloucester cattle, Shetland sheep, the breeding status of the flocks at NAC and at Reading, replenishing the Manx Loghtans and the Whitefaced Woodlands at the NAC, breeding of the Norfolk Horns, and a breeding programme for the Oxford Sandy and Black pig. Most important of all, they turned their minds to the 'Proposed National Organisation' and recommended firstly to continue the Working Party arrangements for the next 18 months or so to allow information and photographs to be collected and compiled into a booklet 'as the document on which the National Organisation might be launched'. In the meantime the rare breeds would be maintained 'rather as a Farm Zoo whereby the NAC would act as a shop-window for the various flocks and herds of rare breeds located in their different parts of the country. It would therefore not be necessary to maintain independent breeding units at Stoneleigh but rather to make contact with those where they existed already.' When the booklet was ready, they would arrange a national press conference and an appeal for funds. It was felt that £5,000 p.a. would be needed to finance the work of a Development Officer whose job would involve maintaining contact with those who already kept rare breeds, ensuring that records were maintained and allowing further development as necessary.

More immediately, it was agreed to prepare a letter for publication in various journals 'to alert the general public about developments in this field' — journals such as *Country Life* and *The Field* as well as the ZSL's Journal and various agricultural and house magazines. It was clearly the intention to broaden the appeal to a very general public indeed, and to escape from the confines of the scientific community which had, for the most part, failed to take as much interest in Zuckerman's vital gene bank idea as had been hoped over the years. Another meeting of the RBWP was arranged for June 1970, at Reading.

The general public did indeed react. Extant correspondence shows interest from, for example, a Mr B.N. Wright describing in *Farmers Weekly* the 1937 Smithfield Show catalogue, which mentioned entries of four Lincolnshire Curly Coat, six Cumberland, four National Long White Lop-eared and two Dorset Gold Tip pigs — three of which were extinct by the late 1960s. A land agent in Glamorgan mentioned that a client had some Whitefaced Woodland sheep; a specialist poultry breeder offered a pair of St Kilda sheep (a type later known as the Hebridean), and a few months later Mr P.S. Turner from Longacre, Norfolk, wrote to say that he had some St Kilda sheep. Ray Carter was able to give him the names and addresses of other breeders, so that by then the NAC was beginning to act as a clearing house and breeders' information bank. Others wrote expressing general interest in the conservation of rare breeds.

Roger Ewbank (who later became director of the Universities Federation for Animal Welfare) wrote from the University of Liverpool to the RASE's veterinary adviser, Captain H.W. Dawes, asking whether the RASE appreciated the risk in getting together a collection of 'rare and unique breeds' in one place, should foot-and-mouth disease strike. This intelligent insight into the problem of low surviving numbers, which would always be at risk to environmental factors, would become the main point of the eventual RBST's first major project: the purchase of an island in order to divide the last remaining flock of North Ronaldsay sheep.

Mrs E.R. (Ann) Wheatley-Hubbard had become chairman of the RASE's Sheep working party, a role in which she was deeply involved with the gene bank animals as her working party supervised the management of the rare breeds at Stoneleigh. She remarked to Stanley (March 1970) on the need for practical farmers on the Working Party as well as geneticists. He explained the background to the rare breeds exercise and told her that the Working Party would continue for about 18 months while the NAC acted as a shop window for the breeds, and that the eventual aim was to form a national organisation for the rare breeds.

Ann Wheatley-Hubbard became one of the most effective people in the rare breed movement and would be one of the Trust's most efficient chairmen. She attended the third meeting of the Working Party on 4 June 1970, at Reading University's farm at Sonning. This meeting was also Joe Henson's first. The meeting agreed to work towards a national organisation supported by public subscription: Christopher Dadd undertook the task of preparing detailed proposals. Also, through Mrs Wheatley-Hubbard and the RASE, a sum of £50 towards a national survey of the breeds was made available.

The summer survey

The survey — at last! It was at least a decade too late for some but without it not much could be done realistically about the rare breeds. Quite separately, in January 1970, Alan Marsden (who kept a flock of Portland sheep) had written to MAFF asking if the Ministry would be 'willing to hear the case briefly for the preservation of scarce breeds with a view to sponsoring my making a survey of the present position. This appears to be unrecorded. It is a necessary first step to setting up a conservation project for which I believe there is sufficient support and which I intend to do.' His offer was passed to the RASE but not taken up, it seems.

Bowman suggested at the June meeting that his students could carry out a survey during the summer vacation. Ryder-Davies generously submitted a long document from his private survey as the basis of the new one. The two students who travelled the country that summer hunting for rare breeds were Mary Underwood and Charles Aindow, and the latter still has some cheerful memories of it all. Bowman gave them lists of names and addresses as starting points and also sent a circular to breed societies for more contacts, and from then on each breeder or owner they met could tell them of others.

One of Aindow's most vivid memories of that summer was a visit to the Dowdeswells and their herd of Gloucester cattle. It happened that Chris Rogers, who had been farm manager at Stoneleigh in the early 1960s, was in the later years of the decade a Ministry livestock health advisory officer in Gloucestershire. He knew the Dowdeswell sisters, who were becoming increasingly frail with age and found it difficult to handle their only remaining Gloucester bull. Rogers recognised that the breed as a whole was almost extinct and made a special plea to the Ministry that semen should be collected from the bull for future AI in the herd, to avoid crossbreeding. Fortunately, his senior happened to be Bill Longrigg who was at the time attending meetings of the Working Party, and it was agreed that a private licence should be given for the Milk Marketing Board to collect and store the bull's semen for AI, so that the relative purity of the Wick Court herd could be continued. Rogers is now Meat Trades Advisor with the MAFF and remains a useful link for the Rare Breeds Survival Trust.

Meanwhile the Reading students visited herds and flocks all over the country and completed their survey. How accurate was it all? Well, time, knowledge and experience were limited: they just did the best that they could. Aindow, today an arable farmer in Lancashire, based his thesis on the survey. He still has the photographs he took that summer as well as his good memories.

The preliminary results were circulated to the Working Party in October that year and would be pulled together and published by Bowman and Aindow in 1973. While there were inevitably many gaps and inaccuracies, given the short time available to the students, the survey did serve to highlight some of the breeds in need of preservation. It also drew attention to the names of people who had been keeping rare breeds. For example Alan Marsden was the owner of two male and eight female Portland sheep; Mr Denys H. Stubbs was another Portland owner, from Lichfield in Staffordshire, with eight females, a ram and a ram lamb — he would later become an RBST Council member. Harpur Crewe of Calke Abbey had about 40 Portlands (including two rams) and this flock would become part of the Trust's history. The survey noted owners of White Park cattle. Apart from the bull, four cows and three heifer calves at the NAC, there was the Dynevor herd consisting of a bull, eleven cows and assorted youngstock, and the herd at Hedenham Hall in Suffolk belonging to Lord Ferrers. Gathered from Whipsnade and Woburn, his herd consisted of one bull, 13 cows, ten young heifers and some calves. Gloucester cattle owners included Miss Salter (who later gave some of her cattle to the Trust). Major Savage at Battle, Sussex, seemed to have some Glamorgans. Richard Body, M.P., owned a herd of Berkshire pigs, and Geoffrey Cloke of Solihull, Warwickshire a herd of British Lops (Cloke would become a major stalwart of the Trust), and of course Mrs Wheatley-Hubbard a substantial herd of Tamworths — more than a third of the entire breed in the UK.

But Ann Wheatley-Hubbard was still worried about the general attitude to the rare breeds. She wrote to Dadd in July 1970:

> I am slightly concerned whether we are only maintaining a zoo, with or without the co-operation of other interested breeders, or whether we are intending to see if any of these breeds have any commercial use for the future. It seems so much more fashionable at the moment to import breeds from all over the world and test them and I feel that we may well be missing out on the advantages of some of our own rare or more unfashionable breeds in this country. I feel that the efforts of our committee will not really be justified unless we can become something more than just a preservation society.

That letter sets out the principles which would guide the Trust in due course, and again marks a clear move away from the zoological towards the agricultural value of the animals.

The rare breeds were part of the Royal Show in July 1970. The public began to come alive to the breeds and by September Sir Dudley Forwood had found a donor willing to give the magnificent sum of one thousand pounds towards (specifically) the rare breeds. Forwood was not very keen on a suggestion that foreign cattle should be added to the collection; he felt that it would be better to stick to British breeds 'and not turn it into an international zoo for foreign competitors'. It is a principle to which he remains loyal to this day.

Ann Durham, who had been with the RASE for some time, became Dadd's secretary during 1970 and began to attend and minute meetings of the Working Party, though it was only one of numerous other committees she serviced and she never felt it was really under the umbrella of the RASE. She would continue to handle the RBWP minuting until, with some relief, she handed over the papers to Peter Hunt in 1972.

The link between the RASE and the Trust has remained strong to the present day. The early link through Forwood and Dadd would continue through several chairmen of the Trust: Wheatley-Hubbard (Trustee of the RASE), Richard Cooper (RASE Council member) and two RASE vice presidents, Michael Rosenberg and Dudley Reeves. The relationship has been reinforced by several others who have played a prominent role in both organisations.

The private collections

As well as the corporate agricultural side represented by the RASE and its National Agricultural Centre, the strands in the web that would form the RBST now included the zoological interest (ZSL, Whipsnade, Wellcome Institute and individual researchers elsewhere), the academic (the University of Reading in particular) and several individuals with a practical interest in livestock breeds. Among the latter were those who collected rare breeds for their own pleasure, whether to decorate parkland or to rear on a smallholding or as a sideline on a farm, and those who saw that there was a growing leisure industry in which farm livestock could play a part. These were the people who would open their premises to the public, perhaps in the interests of education or as working environments for those with disabilities, or to attract paying visitors whose donations would help to maintain the livestock. The disappearing breeds were not only rare enough to excite curiosity and nostalgia; many of them were also striking in appearance, with dramatic horns or unusual colours in a countryside which by the post-war years was dominated by black-and-white dairy cows and long, lean, white pigs reared indoors.

Corporate bodies, especially city councils with spare land, also saw the possibilities offered by rare breeds; Portsmouth was one of the first. Among individuals, by 1967 Rex Woods of Ashe House, Musbury, in Devon, had already established a rare breeds centre. Joe Henson was finally able to open his Cotswold Farm Park at Guiting Power in 1971.

A national organisation

The fourth Working Party meeting, its agenda issued on RASE paper, took place on 27 October 1970 at Bemborough Farm, with Ann Wheatley-Hubbard taking the chair. Dadd's detailed proposals for a national organisation were considered and even the title of his paper provoked discussion. He called it 'Lesser/minor/special breeds group/ organisation/ association'. During the discussion it soon emerged that the word 'rare' was an important one, and also 'survival'.

The paper outlined a set of objectives for an organisation which would be concerned with 'breeds of domesticated livestock which, in the absence of commercial demand, risk elimination'. Under 'Objectives' Dadd listed:

The Organisation will:
1. Maintain contact with breeders and keep a register of animals.
2. Provide liaison between breeders as may be appropriate (viz exchange of sires).
3. Arrange for the provision of technical and breeding policy advice when requested to do so.

4. Formulate and make known the Organisation's views on the breeds with which it is concerned.
5. Make arrangements for the preservation of breeds and breeding groups having characteristics thought to be worthy of preservation.

It has been agreed that breeds, numerically small, may justify preservation for one or more of the following reasons:

1. Historic reasons (viz Chillingham Cattle).
2. Zoological interest.
3. Preservation of genes of special or potential value commercially, for instance in hybridization work.

It has also been agreed that the breeds with which the Organisation may be concerned can be divided into three main groups:

1. Breeds with very low numbers of breeding females, say less than 100, to be known as *rare breeds*.
2. Breeds which do not appear to have specific commercial value (as pure breeds) and are tending to decrease in numbers of pure stock, to be known as *minor breeds*.
3. Breeds which may be of commercial importance outside the UK but of interest largely or wholly for periodic or limited hybridization programmes in the UK to be known as *special breeds*.

He then considered details of the organisation itself. It would consist of a Council 'which shall appoint a Scientific (or Livestock Advisory) Committee' and any other committees or working groups as it deemed necessary. The Council would consist of representatives appointed by certain national organisations (which were listed separately — national and regional agricultural societies, MAFF, ARC, DAFS, DANI, MMB, MLC, NFU, CLA and FPS); not more than ten people representing farm or zoological parks 'who may be corporate members at any one time'; not more than five people representing individual members of the Organisation 'who may be breeders or others interested in the work of the Organisation'; and not more than four scientists or technologists, acting in an honorary and advisory capacity. The Scientific Committee could appoint additional people to give scientific or professional advice.

The organisation would be funded 'from subscriptions and from such grants or gifts as may become available'. The national organisations already listed would pay £250 p.a.; Corporate Members £40 p.a.; and individual members £2.50 p.a. It was expected that those taking part in the Organisation's affairs (i.e. Council or committees) would do so on a voluntary and unpaid basis. However, one of the national organisations should provide general administration at a negotiable annual fee, and a technically qualified person would be employed (by that national organisation) whose time would be spent wholly on the new organisation's affairs until its Council decided otherwise.

The early activities and aims of the new organisation would include agreement about breeds in each group; discussions with any existing organisations which might have an interest in those breeds and with breeders (the intention being to draw up a definitive statement of the present status of each breed); definition of the known qualities of each breed and consideration of the institution of studies by qualified people with a view to 'defining in scientific terms suspected especial genetic qualities'; definition of the practicability and objectives for maintaining individual breeds within the organisation's ambit; preparation and publication of an illustrated booklet, for sale, with the two-fold aim of making the organisation's objects known and bringing in revenue; the undertaking of a comprehensive study on how the initial objectives might be achieved and to formulate a national long-term policy; and finally, to circulate a news sheet to all members, two or four times a year.

Dadd indicated that the facilities of his NAC would be made available so that up to six animals of each breed could be displayed to farmers and the general public. However, he pointed out that 'it is not practical to expect that breeding flocks and herds can be maintained at Stoneleigh but rather to maintain contact with Breeders around the Country. The replacement of animals therefore will come from individual breeders.'

Next he proposed a preliminary administrative budget, for perhaps the first three or four years, based on £3,500 for the first year and allowing thereafter annual increases of 10% for inflation and 10% for any increase in activities and salaries. The first-year budget included an administrative fee to the national organisation which undertook to provide office and clerical support (£600); the salary of a Technical Officer (£1,850); the officer's expenses and direct office costs such as telephone, postage and stationery (£1,500) and a contingency of £550. To produce a viable income, Dadd said that there would need to be about 12 member national organisations, 30 farm park members etc and 100 individual members at the subscriptions already set out.

Christopher Dadd was a man of considerable foresight. He saw himself as a communicator and he was an excellent one. Looking back to those early days (five months before he died in 1992), he remembered having been attracted by the concept of a gene bank right from the start: the idea of maintaining genes, especially from historically important breeds, for introduction in the future if they had a value, seemed eminently sensible to him: 'We should not let them disappear if they have potential value.' And so he began communicating — talking to all sorts of people in different fields, listening, investigating the Whipsnade animals before recommending to the RASE that they should become involved. Then he went to Ann Wheatley-Hubbard, who immediately recognised the value of the gene bank idea and encouraged RASE participation. Dadd began to think of other people who might be drawn into the scheme and developed his ideas about saving rare breeds: the task should be carried out properly under the auspices of a charitable organisation 'to which people could subscribe their energies, their cash and their interest to make it all work.' The NAC took on its gene bank sheep flocks from Whipsnade and, looking around Stoneleigh, Dadd saw enough space for some of the White Park cattle as well.

Although at that time Dadd saw a relevant role for the NAC in the preservation of genes, he did not expect it to be a permanent one. He always intended to disperse the care of the animals, admitting that rare breeds were 'not really what we wanted to communicate to normal farmers'. After all, the NAC was intended to be bang up to date, not looking back to the good old days. His personal philosophy in establishing some of the livestock units at Stoneleigh and getting the agricultural industry involved was to find out where there was a common overlap of interest.

The central core of the new rare breeds organisation for which he had produced the outline was quite clear and mutually agreed from the start, though there was plenty of discussion when they started to write down the objectives — and it was discussion, not argument. Dadd was critical, though not aloud, that steps were not taken at once to assess the value of certain characteristics of the breeds they were trying to save, but it was soon apparent that the new organisation was going to be successful in saving those breeds. It acted as a focus for people who had been interested anyway for years in their local breeds — animals kept for traditional reasons, because grandfather had, by people who no longer bothered to register or record them. Those were just the sort of livestock that the new organisation needed to find, tucked away in forgotten corners, and the timing was good: the public was well disposed to the idea of preserving its heritage.

Yes, it was all fine in theory but, like the ZSL, the RASE simply did not have the money to take on the full burden of rescuing the nation's livestock heritage. They needed to create a body which could not only find the technical expertise to save the breeds but could also generate enough public interest to bring in the money to do so.

As Bill Stanley put it recently, from the RASE point of view: 'The main reasons for the decision to form the Trust were that we ran the risk of losing genes which might be invaluable in some future breeding programme — and those breeds were also part of our heritage. But I am sure the overriding reason was that if they were lost we would have been responsible, and that was unthinkable.' Stanley, after chairing the Working Party quietly through the stages that would lead up to the formation of the Trust, stepped back in 1972: he was fully involved in the development of the demonstration units at the NAC and had also been asked to form and chair an East Midlands Regional Panel to replace the county agricultural executive committees. However, he had helped to establish the foundation and was able to hand over the project with confidence in its future development.

CHAPTER TWO

FORMATION OF THE TRUST

TAMWORTH SOW

HEBRIDEAN

CHAPTER 2:
FORMATION OF THE TRUST

THE TASK CONFERENCE

It was under the title Rare Breeds Survival Task that it was decided to call a conference to promote the rare breeds movement, under the auspices of the RASE, at Stoneleigh on 15 October 1971, bringing together anybody with an interest in rare breeds to consider their future.

The chairman of that conference was Sir Dudley Forwood, Bt, who was an important link at this stage in that he was on the Council of the ZSL as well as being deeply involved with the RASE. He was elected as a vice president of the RASE in 1970 and would become Honorary Show Director for the Royal from 1973 to 1977. He recalls suggesting to Christopher Dadd that a meeting should be convened for all those who thought it worthwhile to conserve breeds which were disappearing because they were no longer economically viable. Dadd and the rare breeds Working Party agreed with the idea, and it was left to Dadd to make the necessary arrangements. Henson promised to help by supplying some animals.

Sir Dudley chaired the Rare Breeds Survival Task Conference that October in his capacity as a ZSL Council member and there was a far more positive response than he had even hoped for: there were 72 delegates. The stated aim of the day was to discuss whether a 'concerted effort' should be made to preserve up to two dozen British breeds of cattle, sheep and pigs which might be in danger, to discover precisely what their situation was anyway, and to discuss what, if anything, could and should be done about them.

The speakers are all familiar names in the history of the Trust. The first paper was by Bill Longrigg, chief livestock adviser of ADAS, on 'Early British breeds and causes of decline'; the second was by Dr Peter Jewell, of the Department of Zoology, University College, London, on 'The case for preservation' — about the lessons to be learned from the disbanding of the Whipsnade gene bank and pointing out that where the animals were now on public display they were an unqualified success; the third, by Professor John Bowman of Reading University, was about 'Rare breeds — their position today' and the fourth was 'The origin and history of British breeds of sheep' by Dr Michael Ryder of the Animal Breeding Research Organisation (ABRO). After lunch there was a demonstration of rare breeds provided by Joe Henson and by the Longhorn Cattle Society: they included Longhorn and White Park cattle; Manx Loghtan, White-faced Woodland, Norfolk Horn, Cotswold, Hebridean, North Ronaldsay, Jacob, Soay and Portland sheep; and a little group of crossbred Oxford Sandy & Black x Tamworth pigs which, ironically, were the last of their line and had been on their way to the butchers. Photographs of the animals would appear in the November issue of *NAC News* under the heading, 'Ancient breeds collect at Stoneleigh.'

Christopher Dadd opened the afternoon session with 'A case history', talking about the origin and use of White Park cattle over the centuries and why they had declined. Then Ann Wheatley-Hubbard led a discussion — the most important part of the Conference and fortunately tape-recorded and subsequently transcribed. Contributors to that discussion included Keith Cook of the Milk Marketing Board, as curator of the Board's well established 'semen bank of genetic variation'; the MMB would in due course co-operate with the RBST, through Cook, and would also help the Trust to gain grants for its semen bank from the Balerno Trust. Rex Hudson, a council member of the British Waterfowl Association, spoke on the need to preserve poultry. Geoffrey Purefoy spoke as a prominent breeder of Jacob sheep. Stephen Lance spoke on the possibility of establishing a small menagerie of livestock to be made available for shows; he would later become responsible for co-ordinating rare breed displays at shows, in particular the Surrey County and eventually the Hyde Park event in 1989. Finally Bill Stanley, as chairman of the 'RASE Rare Breeds Survival Task Project', offered a vote of thanks.

The Conference had also received the preliminary results of the survey carried out during the summer of 1970 by Aindow and Underwood from Reading, and this led to more names and information being offered. As a result of the event, 24 people made donations towards the formation of the new organisation and more than 60 people expressed serious interest in it, with another 40 or so writing after the event to state their wish to become involved.

As well as expressing enthusiasm for the general concept, some people raised questions and fears, and indeed some were against the whole idea of an organisation: they felt (and many continued to feel) that their breeds would become associated with a somewhat eccentric set-up involved with 'useless', non-commercial breeds and that this would rebound on their own commercial animals. Among them was John Brigg, secretary of the Longhorn cattle society, who favoured keeping rare breeds commercially rather than putting them in 'zoos' or farm parks.

The most important outcome of the Conference was enthusiastic support for the creation of an 'organisation for the preservation of rare breeds', with Dadd acting as Convenor at the NAC. It was decided, by a unanimous vote during the general discussion, that the organisation should be a trust.

*

Jewell's paper was subsequently published in the *Veterinary Record* (13 November 1971, pp 524—7) 'in view of its topical interest for veterinarians' — which added another strand to the movement. Jewell, opening his paper by saying that the idea of conserving rare *species* had become a familiar one, then built up his case for the conservation of rare *breeds* of domestic animals, drawing parallels between breeds in domesticity and the varieties or races in wild species. 'I think,' he said, 'their value lies in several directions: they have value for scientific research, but in addition, today, another value is gaining ascendancy and that is their value as a "living museum", to be enjoyed in recreation and used for education.' He pointed out that even Darwin, after he had built up his ideas on the evolution and origin of species, turned his mind to variation under domestication but that thereafter zoologists had tended to leave the study of domestic animals to agriculturists. Zuckerman's gene bank had swung the balance again towards zoologists, and the new rare breeds movement could now gain by co-operation from both sides. But Jewell's vision was a wider one: it took in historical and archaeological research on the one hand, and on the other a quite unscientific reason why the breeds should be conserved: a reason which was 'cultural in the very broadest sense' — the educational value of collections put on display to the general public.

> Townspeople, it seems, are keen to seek diversion in looking at animals and the recreational value of keeping all breeds in being is, I think, something that will grow enormously in the next few years. To consider the broader question of recreation in the countryside we already see how problems are arising because some Nature Reserves and places of outstanding natural beauty have become so popular that they are suffering damage. It has become of increasing importance to offer other kinds of diversion. It seems to me that the farm park will provide exactly one of these diversions.

He wound up his talk by saying that he had quite deliberately broadened the background against which decisions to preserve must be taken and said that, in the last resort, the only cogent argument was based on the broad cultural, scientific and educational value of the breeds. In other words, he was suggesting what the subsequent development of the RBST would mirror: rare breeds appeal not only to the scientist and the pedigree breeder but also to the public at large for, frankly, sentimental reasons. And where is the harm in that? The farm animals are, after all, our living heritage, and we have a duty to future generations to preserve their potential to meet unknown future needs.

Idwal Rowlands, remembering Jewell's paper more than two decades later, remarked recently that of course nobody ever saw the Gene Bank animals at Whipsnade and they were almost forgotten for that reason. And it didn't help that they were largely left to their own devices 'to lamb on the downs — the coldest place in England!'

*

Michael Ryder's report on the Task Conference was published in *Nature* in November 1971 and it was this which had elicited the quick response (already cited) from Zuckerman in the same journal, correcting one or two points of record about the Gene Bank at Whipsnade.

FORMING THE TRUST

Bill Stanley's Working Party met four or five times during 1972, mainly at the RASE's London offices in Belgrave Square, with the aim of launching the new organisation in 1973. They discussed detailed ideas about the formation of a trust, drafting its deeds and suggesting possible presidents and candidates for what would eventually become a Council.

They had a few problems in the drafting of the deed: the Charities Commission did not seem to like the phraseology of the legal documents produced by a London lawyer on the group's behalf. Fortunately Henson's personal solicitor came up with the name of an expert in setting up charities. Peter Hunt, previously with Dr Barnardo's Homes, had become a self-employed management consultant (trading as Peter F. Hunt & Associates) in 1971, working from his home in Abbots Langley largely on charities, with a special interest in the relationship between their professionals and their volunteers (always a potentially volatile situation). Looking back to those days, he now admits to having no knowledge of rare breeds then, though he had childhood memories of Shire horses drawing coal carts along the streets of Watford. (His work with the Trust later inspired him to keep rare poultry breeds, until the local foxes became too successful in culling them.)

Hunt initially met Henson at the Cotswold Farm Park and they spent the day discussing the possibilities. Then the two men had a meeting with Christopher Dadd and Hunt subsequently drew up his proposals for the Working Party in a paper dated 18 December 1972. His opening remark was that, before major fund-raising could take place, there would have to be a period of establishment during which 'membership must be built up and needs identified'. Hunt, acting as Secretary to the Trust, would deal with membership records, accounts, servicing meetings of the trustees, general administration, fund-raising and public relations, allowing the Trust to grow gradually or, as he put it, 'for its involvement in fund-raising to grow in parallel with the fulfilment of its prime purposes, i.e. the assistance of survival of rare breeds'. During this initial period, Hunt's consultancy would approach grant-giving trusts ('especially those known to favour the provision of starter finance and support of environmental causes') and follow up individuals who had already expressed some interest in the aims of the Trust. He foresaw two main areas of financial need: (a) administration of the Trust, 'i.e. salary of field officer, membership records, newsletters, conferences etc.' and (b) emergency situations requiring capital input, possibly at short notice — for example, to save a herd about to be slaughtered, and possibly other 'highly emotive' situations requiring an urgency that demanded a 'different level of activity' from normal Trust administration. Acknowledging the need to keep the Trust's initial financial commitment as low as possible, Hunt carefully limited the scope of his own involvement and asked for a fee of a mere £750 p.a., which would also cover the use of his own office and staff. When someone at the meeting commented that they had no money to pay him even that, he responded that if he could not raise enough to pay himself, he was not much of a fund-raiser, and that he thought the whole project so interesting and worthwhile that he would in effect start working for nothing. He knew which grant-giving bodies to target. Hunt was immediately asked to become secretary and fund-raiser of the Trust.

Hunt's first task was to gain the acceptance of the Trust by the Charities Commission. He took a look at the deeds which had already been drawn up and saw immediately that they needed radical re-drafting, based on his specialist knowledge of charities. One of his main decisions was to limit the number of trustees in the new holding trust deed, and these were named as Cranbrook, Forwood and Henson. Lord Cranbrook was ideal: he had chaired the ZSL's Collections Policy committee in 1959, working in liaison with the newly formed Breeding Policy committee; he had founded the Mammal Society and was closely involved with the Nature Conservancy; and he had already agreed to become the new Trust's first president.

At the time the Charity Commission had a venerable set of objectives which were deemed to be acceptable for charities, and the Trust therefore emphasised its educational and scientific aims.

Right from the start, they had decided that the next stage would be a company limited by guarantee (and not having a share capital), and a membership-based organisation. Hunt's view was that, as a charity which would require on-going financing, it needed to develop a broadly based and involved membership: his phrase, learnt at Barnardos, was that 40 sixpences are worth more than a pound — if you lose the one pound donor, you lose everything, but if you lose one of the sixpenny donors you still have the rest of them. He thought that the membership should include not only those with a direct interest in breeding but also a wider population who could see the need and justification for the Trust's work and were persuaded that they were not just a bunch of cranks. Although some did not and still do not agree with the idea of a large membership, everyone was very aware of the risks of that 'eccentrics' image.

There were other small developments in the early part of 1973. Mrs Barbara Platt, for example, of Bank Farmhouse, Down Holland, Ormskirk, offered Dadd two ewe lambs and a ram of the Manx Loghtan breed: she was a Manxwoman by birth and would become an important part of the Trust's Combined Flock Book team in due course. On the first day of the year, Alastair Dymond became farm manager at NAC: he would in due course have an influential position in the Trust. Also early in the year, the Duke of Wellington was interested in opening a country park at Stratfield Saye: he would become president of the Trust later.

The Working Party meeting on 15 February 1973 was again held at Belgrave Square, with the agenda still on RASE headed paper under the title Rare Breeds Survival Trust. At this stage a Steering Group took over from the Working Party with the aim of putting into effect all that the Working Party had prepared. Cranbrook was there as president and Henson took the chair for the first time; all the old hands were present – Dadd, Forwood, Rowlands and Wheatley-Hubbard, along with half a dozen others. The main item for discussion was the temporary trust deed, intended to be in place until the memorandum and articles of association could be finalised. Hunt's appointment as Acting Secretary to the Trust was made official, on a very modest financial basis and with no security of tenure, and he was authorised to enter into negotiations with the Charity Commissioners. It was, inter alia, agreed that initially the Trust should be concerned solely with British rare breeds of domestic livestock.

Hunt worked very fast in dealing with the Charity Commissioners. In what was for him a record time of just a fortnight, he obtained their approval of the newly drafted Trust deed. In March the Rare Breeds Survival Trust was officially registered as charity number 269442. By April Hunt had set up a bank account for the Trust and wrote to Dadd requesting 'a cheque for the donations the Royal Society are holding on behalf of the Trust. I have so far received just over £200 in donations and membership fees, which I have banked.' He mentioned to Dadd an enquiry from Michael Ann, of Drusillas (an animal collection in Sussex), asking for help in devising a suitable stand for use at the South of England show.

Hunt's speed and efficiency were impressive. His expertise and experience in charity work were crucial to the fledgling Trust. He insisted that the way forward was to form the charity first and then to raise the funds to do so — people would be far more prepared to donate to an existing charity than to some vague ideal of the future. It was on 19 May 1973 that the press notices proclaiming the formation of the Trust were issued.

A philosophy for rare breeds

The philosophy of the Trust was formalised eventually in the full Memorandum & Articles of Association of the limited company formed in March 1975; but was already the impetus for the new charity in 1973. The objectives were:

> For the benefit of the public to ensure the preservation of breeds and breeding groups of domestic livestock of importance in the promotion of agriculture being numerically small and having characteristics worthy of preservation in the interest of zoological research and education to ensure the preservation of genes of special or potential value in hybridisation and other work.

The word 'preservation' is much in evidence. Perhaps at that stage it was used a little loosely: preservation tends to restrict, casting the subject in stone for ever, keeping it exactly as it is at that moment. But the real role of the Trust would be in conservation, in which it is accepted that life does not stand still, that there is dynamic evolution even in rare breeds. The semen bank, however, would be for the preservation of certain genes literally frozen from a moment in time.

The phrase 'for the benefit of the public' was essential to achieve charitable status, and also the use of words like 'education' and 'research'.

Then there is the phrase 'breeds and breeding groups of domestic livestock': it does not include wild animals, though it could be stretched to feral livestock (a point which would be argued, forcefully, by Peter Jewell and his wife Juliet Clutton-Brock for several years before the Trust could be persuaded to include a category for ferals). It referred to 'numerically small' populations and to those 'having characteristics worthy of preservation in the interests of zoological research and education' — both concepts still respected in the Trust today, the latter by means of the Show Demonstration programme and other means of bringing the breeds to the attention of the general public. Then there is the phrase 'the preservation of genes of special or potential value' — and that word 'special' is important: there are some breeds today which are 'numerically small' but which the Trust does not conserve because it does not consider their genes to be 'special'. For example, if two breeds are crossed to create a third, the latter is not 'special' while it is a recent creation because it has nothing new to offer. It could be argued that this is the case with 'reconstituted' breeds or with brand new breeds created by combining several others, such as the Colbred or the British Milksheep, even if they should happen to be numerically small. But then there is the phrase 'potential value in hybridisation and other work': that, clearly, brings in the importance of the breeds to agriculture — the genetics angle of using the rare breeds quite specifically and deliberately at some stage for cross-breeding in a productive way.

That long sentence defining the Trust's objectives holds true some twenty years later. There are alternatives. For example, the Trust could work with 'groups' rather than breeds of livestock — groups of a defined type or appearance — or with specific genes, with DNA, with a whole range of genetic material. It could work with rare animals, all put together in a large gene pool as one big population, a gene soup. But it decided against all those approaches: its philosophy and priority remain, wherever possible, to work through recognised and established breeds of livestock. In some cases, however, there were groups which did not have a breed society when the Trust began its work but which were proved to have been a self-contained population for long enough to be considered a breed: for example, the St. Kilda sheep which later came to be known as the Hebridean and for which the Trust itself acted as the breed society.

Next it is worth considering why there should be any conservation of any breed anyway. There are perhaps four separate reasons. The first is as insurance against what might happen in the future: if a breed is lost, it might have had characteristics which could have been of enormous value at some time in the future. Secondly, some of the breeds might well have current commercial value; it is simply that their potential has not been evaluated — the White Park cattle and Hebridean sheep have proved their potential under the Trust's auspices, for example. Thirdly, some breeds are of peculiar scientific value for their special characteristics — the North Ronaldsay sheep is a good example, with its physiological adaptation to a diet of seaweed. Perhaps some of the other primitive sheep will prove to have carcases high in polyunsaturated fats.

Finally, there is the sentimental angle of heritage and history — the incentive for many of those involved in the formation of the Trust. Ann Wheatley-Hubbard, for one, distinctly remembers the strong feeling that the motivation was for purely historic reasons: they had already managed to lose the Lincoln Curly Coat pig, for example, even while they were all talking about the rare breeds concept, and they were very conscious that it was 'part of the history of the country and therefore it was essential to preserve these breeds.' She told Lord Montagu, when he was creating his motor museum: 'You can always make a car, but you can't make a breed if you have lost the genes.' The prime aim was to preserve the heritage by preserving the genes, regardless of any 'use' those genes might have.

The Advisory Committee

It was an invigorating time for all those involved with the rare breeds movement. In May 1973 the three Trustees appointed an Advisory Committee 'to assist them in the formidable tasks that lie ahead'. This committee formed the basis of what would eventually be a Council when the Trust became a limited company two years later. The members during 1973, a mixture of private individuals and those representing various organisations, included in alphabetical order: G.L.H. Alderson, M.D.M. Ann (Drusillas), Professor J.C. Bowman (University of Reading), J. Cator (breeder of British White cattle), J. Cole-Morgan (Agricultural Research Council, to advise on public relations), R.P. Cooper, C.V. Dadd (RASE), Dr J.F.D. Frazer (in attendance), Professor P.A. Jewell (Department of Zoology, Royal Holloway College), Rt Hon Marcus Kimball, MP (though he never actually attended), W.A. Longrigg (MAFF), A.J. Manchester (NPBA), Captain C. Pitman (Fauna Preservation Society, succeeding R.S.R. Fitter), Dr I.W. Rowlands (ZSL) and Mrs E.R. Wheatley-Hubbard. Many of them had already been influential in getting the rare breeds movement on its feet.

Appeals and Funding

Peter Hunt now launched a major appeal to elicit starter finance from the grant-giving trusts and to establish an emergency fund. This first appeal document looked for £30,000, including £5,472 for compiling and maintaining an information bank; £7,625 for a mobile field officer to undertake research, inspect flocks and animals, and provide assistance and advice; £4,150 for promotion etc., giving a total of £17,250 for general running costs. Then came a section for emergency actions: a sum of £2,750 to acquire and maintain for three years a sanctuary for North Ronaldsay sheep. And finally £10,000 was set aside as a reserve fund against future emergencies, including the taking up of any offer of suitable land for other sanctuaries.

The document stated that the formation of the Trust marked the conclusion of the Working Party set up in 1968 by the RASE and ZSL, the purpose of which had been to 'look for ways and means of preserving rare breeds of British farm animals'. It said that an interim trust deed had been agreed with the Charity Commission in March 'in order that there should be no further delay' in what was clearly an urgent need for a permanently established organisation 'that can keep a check on known rare breeds so that the need for drastic action can be avoided. Hence the Trust was born.' (The 'drastic action' was the attempt to rescue the Norfolk Horn, whose extinction had been predicted as long ago as 1846 by David Low, when it was already in decline and becoming rare.) The document also stated that the 'long term intention is for the Trust to become a company limited by guarantee when it is hoped to have a council of management elected by the members of the Trust who will be drawn from all walks of life but particularly those concerned with the land and the care of animals.' (The intention here was to get members actively involved.) It set out the case for the preservation of rare breeds, ending with a rousing declaration: 'Finally and by far the most important reasons for many people are those of sentiment and tradition. Future generations will want to see living examples of the animals which are part of their heritage. A stuffed dodo does not compensate for the living and walking bird.' It explained how the Trust would meet its objectives — acting as an information bank, bringing together owners to encourage and facilitate the breeding of the animals (and if necessary giving grants towards the purchase of animals or even purchasing them for the Trust and, if there was no other way to preserve the breed, the Trust would own and maintain breeding stock), arranging conferences and exhibitions, holding frozen semen for AI, co-operating with breed societies and similar organisations for the best use of available resources, ensuring that the gene bank was used 'for the improvement of breeds currently in use and for the development of new breeds to meet the problems brought about by changing economic and environmental circumstances', and giving the public opportunities to view the animals and learn about them — the Trust, in the interests of education, would encourage the setting up of farm parks and static exhibitions 'of a high standard', it would promote media interest, and it would publish 'scientific papers and leaflets of general interest'.

Hunt's paper then set out the breeds thought to be in the most immediate danger (sheep: Black Welsh Mountain, Cotswold, Dartmoor White-faced (horned), Lleyn, Manx Loghtan, Norfolk Horn, North Ronaldsay (Orkney), Portland, Rhiw, Soay, St Kilda, 'White-faced Woodland and Limestone', Wiltshire Horn; cattle: 'British White Park', Dexter, Gloucester, Irish Moyle, Longhorn, Shetland; pigs: Berkshire, Gloucester Old Spot, Lincolnshire Curly Coat, Middle White, Oxford Sandy and Black, Tamworth, Ulster White). Purists will notice several anomalies in this early list.

Finally it came to the main point and looked at what it would all cost for the first three-year development programme: updating the 1970 survey, setting up the information bank, establishing field services through a mobile field officer, promoting the needs of rare breeds and establishing its first breeding sanctuary for the North Ronaldsay sheep. The grand total was £30,000. It was hoped that:

> in the long term the costs of running the information bank and providing the field services will be largely met through members subscriptions BUT it is clear that the setting up costs, and the recurrent expenses of education and promotion, together with rescue operations such as that planned for the North Ronaldsay, will require sustained fund-raising. The trustees hope that all who receive this pamphlet will consider helping in this urgent and vital work of preservation. Some will want to be members of the trust because of their own direct involvement with rare breeds. The immediate needs are for *starter finance* — to get the project off the ground, *recurrent grants* — especially for the first three years, and *donations* — especially those given under deed of covenant thereby enabling the trust to have some degree of stable income.

The address at the foot of this first major appeal was that of the Secretary at Ashleigh House, Abbots Langley. The artwork included groups of 'St. Kilda rams', Highland cattle, Soay sheep and the rather crudely drawn head of a White Park, which became the symbol of the Rare Breeds Survival Trust and remains so today, though its representation has become more refined.

*

Cranbrook, at his Redhouse Farm, soon received a letter from Ralph Verney of Claydon House, Bletchley, saying that two of the trusts with which he was connected had been considering the rare breeds appeal and wanted to help; these were the Ernest Cook Trust and the Radclyffe Trust. However, he wrote: 'One point which occurred to both bodies of the trustees was that we would be happier to give greater support if we felt that the composition of the membership of the Trust included rather more expertise in the realm of the geneticist and the biologist.'

IRISH MOILED

GETTING DOWN TO IT

The very first 'emergency' project was an emotive one with great appeal to the general public. It centred around a remnant flock of primitive sheep on the Orkney island of North Ronaldsay. These animals were not only rare and unusual in appearance; they were also extraordinary in that seaweed formed a large part of their diet, a trait which instantly caught the public imagination.

The Trust conceived the idea of buying a whole island and transferring some of the flock to form a separate breeding colony so that there was a safe back-up population should there be a disaster on the original island: an ecological threat such as oil spillage contaminating the seaweed, or a disease which might wipe out the entire single group in such a confined situation. The idea of a haven was the perfect hook on which to hang an appeal in aid of the newly registered charity. (The full story of Linga Holm is told in Chapter 4.)

Meanwhile the Trust settled down to its important work of conserving a gene bank from a wide range of endangered livestock, and generated the publicity that would bring in financial support for its activities, beginning to build up a membership (as a source of rolling funds, it was hoped). By now the main collections were at the NAC (Stoneleigh) and the Cotswold Farm Park, Guiting Power. In July the NAC's shepherd, Wally McCurdie, reported on the status of the rare breeds in his keeping: there were for example 16 female and 11 male Manx Loghtans there, and he noted that 'several of our ewes have white markings' but these were being culled gradually; three of the eight ram lambs had white markings. The Norfolk Horn remained in dire straits. Among the Cotswolds there were 11 ewes, two shearling rams and several lambs; as for the White-faced Woodlands there were five ewes and three rams as well as lambs, but the note 'NBG' against one of the ram lambs suggested it was not highly rated. It was also noted that some of the White-faced Woodlands had been criticised as not being of true type.

It was inevitable that there was some confusion of thinking in these early days, when it was sometimes difficult to differentiate between the breeds or to establish their lineage. For example, the RASE had written to the secretary of the British White cattle society asking for information on owners of 'Chartley' cattle — a double gaff in that 'Chartley' was not a breed, it was simply a herd of the White Park breed, and anyway the White Park was a completely different breed from the polled British White, though its colour pattern was similar. John Cator, a leading British White breeder, did suggest in the early stages of the Trust that the White Park and the British White should be put in the same herdbook, a situation which occurred in North America and led to considerable confusion as the polled British Whites came to be called White Parks there. The breeds are not related.

At this stage the NAC farm manager, Alastair Dymond, suggested that there should be a rare breeds display on weekend open days after the Royal Show and he sought support from within RASE for the necessary fencing and large, attractive signboards with pictures of the breeds and details about their background. However, his approach to Jef Aartse-Tuyn (whose role centred on communications and publicity) was rebuffed: the latter held the view that 'the rare breeds unit must promote itself. I am providing you with an audience. Should I also foot the bill for presentation?' That was perhaps a shame: the rare breeds needed all the support they could muster at this delicate stage. But, then, the Royal had been going through difficult times itself. In 1970/71, for example, many leading exhibitors were withdrawing from the Show and revenue had become a problem. In a major restructuring, the RASE realised that it had to concentrate on marketing itself, and it had to rely for a while on 'very good friends of the Society' to underwrite, for five years, the salary of an officer newly appointed in April 1972 to shake the organisation into good shape: the new chief executive, John D.M. Hearth, who in effect was appointed over the loyal Christopher

Dadd, though the latter remained as Agricultural Director. Hearth did shake things up on the financial, administration and public relations side of things, and not without some procedural hitches. A working party took a good long look at everything and did not report to Council until March, 1974, so that during 1973 everything was uncertain and people were a little edgy about money and about internal structures. Perhaps it was not the ideal moment for looking after a bunch of uncommercial animals; but it was certainly the moment for gradually withdrawing from the rare breeds project now that a new national organisation had been formed to look after them.

A Technical Consultant

The Trust's Advisory Committee met in June, when it considered the possibility of a cattle semen bank (particularly for the White Parks) and agreed a show demonstration programme. At the August meeting it received Lawrence Alderson's proposals about the approval of rare breeds centres, which suggested two types: approved breeding centres, where the actual work of breeding was carefully carried out, and approved survival centres. This would eventually lead to the approved centre scheme.

They also responded to another Alderson paper suggesting that the Trust needed a technical department. The thought was that they might appoint a field officer and the committee defined the qualities it sought. The appointee must have had practical experience with stock, a broad knowledge of farming practice, an interest in zoology, knowledge of rare breeds of domestic livestock, and an ability to communicate and mix with people at all levels. The total salary and expenses for the first year should be no more than £2,000. The appointment would be on a part-time basis, ideally of someone already self-employed who was prepared to devote an average of one or two days a week to the Trust with a view to the post becoming full-time as and when circumstances permitted. The duties were carefully defined:

1. To supervise and complete a detailed survey of rare breeds.
2. To monitor population movements of breeds and give early warning of trouble spots.
3. To maintain liaison with breed societies and other livestock organisations and to provide assistance for rare breed societies.
4. To act as an agency for the location and sale or exchange of breeding stock.
5. To provide advice on breeding, recording, husbandry etc., as required and to maintain pedigree records where possible.
6. To undertake and encourage project development work, e.g. trials, breed assessment etc., and specific major projects.
7. To carry out promotion and public relations work at agricultural shows, talks, radio etc.

A number of nominations had been submitted. On a proposal from Captain Pitman, seconded by Richard Cooper, it was agreed that Alderson met all the requirements and should be interviewed for the post. Alderson had joined the old Working Party after being approached by Mrs Wheatley-Hubbard following the meeting held in October 1970.

In October 1973 the Trust issued a press notice. 'Rare Breeds Survival Trust appoints first technical consultant: Mr G.L.H. Alderson has been appointed as the first technical consultant to the Trust.' It noted that he was living in Eastrip House, Colerne, Wiltshire; he was involved in consultancy work through Livestock Improvement Services and was a director of Countrywide Livestock Ltd and of Prolific Sheep Breeders Ltd.

The letter appointing Alderson was dated 15 September 1973 and the appointment was with effect from 5 October, initially for six months and thereafter at three months' notice on either side, with annual reviews of the duties and conditions of employment.

It was intended to be on a part-time basis but it was hoped that it would progressively develop into a full-time position. The fee for the first year would be £1,500, covering basic remuneration, general out of pocket expenses including secretarial assistance (if any) and local travel and attendance at committee meetings. The average time was expected to be not less than one or two days a week devoted to Trust affairs. Alderson would be answerable directly to the chairman of the Trust, who in turn would be guided by a technical subcommittee. Alderson would attend meetings of that subcommittee and the Advisory Committee, and would liaise and co-operate with Hunt over administration, public relations and fund-raising. There would be extra expenses for travelling which was deemed not to be reasonably met from the annual fee. Alderson wrote his acceptance of the post of Field Officer under these conditions ten days later, but asked that he should be described as Technical Consultant 'in deference to my responsibilities to my other clients'; he was, after all, self-employed. He also took himself off the Advisory Committee, in order to concentrate on the technical work. Now the Trust could really get down to the *practical* side of its conservation work.

Lawrence Alderson had grown up on a Yorkshire farm with pedigree Northern Dairy Shorthorn cattle, Dales ponies and Swaledale sheep. It was in the days when social life for most Pennines farmers revolved around May bank holiday outings and it was on one of these, in the late 1940s, that Alderson visited Chillingham and saw the herd of 'wild' white cattle, which had been reduced to 13 animals by the severe storms of 1947. This early glimpse of a rare breed was one of several apparently random strands that would later intertwine.

Alderson gained his place at Cambridge on the basis of his sporting record, especially rugby: he was expected to win a Blue and did so, not at rugby but as a boxer. In the same year in which he boxed for Cambridge, Lord James Douglas Hamilton boxed for Oxford and that introduced another strand: the Hamiltons owned the Cadzow herd of White Park cattle.

At Cambridge, Alderson read agricultural science and after graduating, became a lecturer in the unusual combination of genetics and business management. He had no direct involvement in farming but, as a geneticist, he had met the recently retired John Hammond at Cambridge, who introduced him to the British Cattle Breeders Club. Hammond had played a major role in saving the remnants of the Norfolk Horn sheep from 1947 onwards.

Gradually, Alderson's farming roots pulled him back and he became a farmer as well as a consultant. He was particularly interested in the genetics of sheep breeding and bought some Jacob sheep from Sir Walter Burrell, an acquaintance of his wife (who was the daughter of an eminent Sussex veterinary surgeon). Thus he became a founder member of the Jacob Sheep Society.

In the late 1960s, Alderson began to develop his British Milksheep as a business, realising that British sheep in general were not profitable. His determination to make farming pay was the crux of his own farming operation, and the commercial side of it grew considerably when he moved to Northumberland.

In 1968, while working as a company consultant, he acquired a new client, Joe Henson, who was embarking on his rare breeds farm park. Coincidentally, Alderson came to know John Riley, the Wiltshire ADAS livestock advisory officer, and, through him, met Ann Wheatley-Hubbard.

Since 1973 Alderson has been closely involved with the Trust, particularly in the formulation of its policy and the development of programmes. In 1990, on the verge of retiring altogether after two decades, he was called back to become its Executive Director. At heart, no doubt that is what he had always wanted.

The Hunt Newsletters

In August 1973 Hunt had produced the Trust's first Newsletter — a somewhat amateur publication, as he is the first to admit. He typed it himself on an IBM Executive typewriter; his wife Moira (an unsung heroine of those early days who worked almost as hard for the Trust as he did) was the cut-and-paster and took the sheets down the lane to be duplicated. They even included a children's section for a while — a little crossword or word-search puzzle. A great deal of Trust work was done from Hunt's home: its paraphernalia spilled from his little office into his dining-room, with a tendency to spread into the lounge. He was still acting as a consultant with several other clients, employing four or five part-time typists who worked from their own homes: his wife would drop dictation tapes around the village in the morning and collect the work in the evening, and she also handled all the filing. It happened that their Abbots Langley home was (and is) a mere hundred yards from the Ovaltine show farm, once famous for its good Jersey herd.

In the early days Hunt found himself constantly on the phone to Henson, or driving over to Guiting Power for consultation through the early stages of the Trust's formation. The main challenge, Hunt decided, was to persuade the geneticists and scientists that a broad base was necessary, and then to build up that base. Fortunately the Council was 'always a fairly harmonious body — if there were any tensions they were underlying rather than overt'. There were no fist-fights over the big RASE boardroom table in London; they were, says Hunt, 'all such nice people'.

Hunt saw himself as a facilitator and understood his role as being to get the Trust off the ground. He had supervised its registration as a charity and felt by then he had fulfilled his function. But when Henson and Dadd remarked jovially that they had no money to pay his bill and so could he undertake some fund-raising for them, he had liked the idea: he thought the Trust could have quite a wide appeal 'to people like myself who will never be involved in detailed conservation but are the sort of wishy-washy liberals who are members of the National Trust, the Waterfowl Trust and the Worldwide Fund for Nature. Animals are always good for the sentimental touch in the charity world.' And so he had accepted the challenge.

In the end he found himself working on behalf of the Trust for years rather than months, though never on more than a part-time basis, never (officially) more than two or two-and-a-half days a week. By 1976 he had taken on a full-time post as Comptroller of Skinners Hall for the Worshipful Company of Skinners (where he still works today); yet he continued as the Trust's secretary for two or three more years, working in that capacity from eleven at night until two in the morning and over the weekends.

In that first newsletter of August 1973, Hunt was able to report that the membership stood at about 300 and included zoologists, biologists, veterinary surgeons, teachers, farmers, and people with simply a general interest. There were also members from America, Australia, Holland, Iran and Swaziland. He listed the grants which had been promised, gave notice of the Trust's first Field Day to be held in October, and reported that in due course the Advisory Committee would become the first Council of a limited company. Its members at that stage represented many different facets, with some people covering more than one area of interest. Five of them described themselves as farmers, either full or part time: Cator and Cooper, Alderson (who was also an international livestock consultant), Henson (also a farm park owner) and Wheatley-Hubbard (also RASE). In addition the RASE was represented by Dadd and Forwood, and the ZSL (which had started it all) by Rowlands. Michael Ann was simply a farm park owner, though Drusillas was really more of a zoo. There was Arthur Manchester of the National Pig Breeders Association and Bill Longrigg of ADAS. On the academic side there were Bowman and Jewell from the universities; there was Cole-Morgan from the Agricultural Research Council. With a wider interest, there were Cranbrook of the Natural Environment Research Council (not forgetting his strong associations with the ZSL) and Pitman of the Fauna Preservation Society. It was quite a rich mixture: zoology, agriculture, natural history, conservation, research, leisure and so on. Many of the original Council members stayed on Council or its various committees for many years. It is to those original stalwarts that the Trust owes its success.

Hunt's first newsletter also referred to the question of farm parks (one which would come up time and time again later on), pointing out that the first three which had come to the Trust's notice were Michael Ann's Drusillas at Alfriston, Sussex; Cotswold Farm Park (Henson and John Neave) at Guiting Power; and the Farway Countryside Park near Colyton in Devon. He also reported that the Trust now had an offer from the Ernest Cook Trust to cover the salary and expenses of a Field Officer for two years.

The date of commencement of Alderson's appointment was also the date on which the Trust held its first members' field day and conference, at the Cotswold Farm Park, and it drew more than 200 people to see the breeds and talk about finances, membership and priorities. (More details of this first event are given in Chapter Six, which also covers some of the other field days and annual general meetings in the Trust's history). Hunt's December newsletter gave a full report and also celebrated the success of Black Welsh Mountain sheep at Smithfield and the fact that, at last, the variously named North Ronaldsay or Orkney or Linga Holm project was well under way: an offer had been made for the island of Linga Holm and the sheep had been purchased. The year of 1973, which had seen the official birth of the Rare Breeds Survival Trust, ended very much on the up-beat.

Hunt's newsletter continued valiantly for a while. In March 1974 it was discussing whether bees should come within the Trust's remit: it was pointed out that the old 'native' bees of Britain had probably been lost early in the 20th century. However, the Trust decided to concentrate deliberately on larger livestock; species such as bees, rabbits and other peripheral livestock never found acceptance.

At the Advisory Committee meeting that March, the newsletter's fate had already been sealed. Alderson put forward a proposal from Michael Rosenberg to produce a monthly journal on behalf of the Trust, and the first issue of *The Ark* was published in May 1974. The same proposal dealt with the establishment of the Combined Flock Book.

First Annual Report

In their first annual report (to 31 March 1974), the Trustees were well pleased with the year. They carefully thanked certain grant-giving trusts for getting the organisation up and running. They mentioned 'several anonymous gifts' and said that they had received donations from a wide variety of people, from old-age pensioners to young children.

Within its first twelve months as a registered charity, the Trust's membership had increased to nearly 500 (including 14 corporate and 47 junior members), which might seem tiny in comparison with today, but by December 1974 they had added another 50%. The Trust was very active indeed now: the years 1974—75 would be momentous ones as its members' and officers' imaginations got to work on every aspect of its task. The Advisory Committee looked at ideas on education; it tackled other organisations and achieved the agreement of the Meat & Livestock Commission to include rare-breed bulls in its performance testing, and of the NPBA to register its pig breeds. It began to gain commercial sponsorship. It developed an ambitious Show Demonstration programme; it supervised the opening of more rare breeds centres and developed a set of criteria for their approval. A new survey was completed and international contacts were being made.

Catching the Media

The useful publicity continued, though it was not free from controversy, which highlighted a divide within the Trust. During the year there was an article in the *Daily Telegraph* colour supplement which delighted many, in that it drew attention to the rare breeds and to the work of the Trust, but dismayed some who considered that its sentimental approach belittled the Trust's serious technical work. This remains something of a dilemma even now: the business of expert breeding and genetic analysis on the one hand, and on the other the popular appeal which can generate the funds needed for that business. In the early days there was certainly a tendency to label the Trust as a group of eccentrics, but many now famous and substantial organisations had the same label in their youth — bodies such as the RSPCA, the RSPB and almost anything concerning itself with animals other than human ones. Yet the label rankled among some of those with professional and academic backgrounds in science and agriculture, and would remain a source of friction, albeit often constructively so.

Publicity is lifeblood for a charity, as the Orkney appeal showed. The Show Demonstration programme ambitiously took in major venues in 1974: the Surrey County, the Royal itself, the Royal Cornwall and Smithfield for example. Alderson published an article in the RASE Journal about rare breeds and the Trust's work.

Towards Independence

It was towards the end of 1974 that the RASE began to pull back more obviously, encouraging the Trust to stand firmly on its own feet under Hunt's administration. Until then the RASE had offered the free use of its own London offices for Trust meetings and had generously given secretarial help as well; now it suggested that there might be a charge for the use of the Belgrave Square meeting room. Its withdrawal was certainly not for lack of interest in the rare breeds movement but Dadd had always been told that, more or less, this rare breeds business was his baby and not really part of the NAC's modern image, and anyway the NAC had its own financial problems. The Trust was beginning to look like a viable concern: it was accumulating funds and the bank account by June 1974 stood at nearly £11,000. Hunt's financial report to the end of September showed a total income of £12,356, including donations amounting to £9,722; after expenditures there was a balance of some £7,000 put away as a reserve to meet the expected costs of the Linga Holm project plus some £1,700 in the bank. The projection for 1974/5 was for a total income of £13,806 (including nearly £7,000 from the Linga Holm appeal) which, after deducting some £6,000 for anticipated expenditure, would give a reserve of £4,500 and a positive balance of some £3,400. However, the estimate for 1975/6, by which time the Linga Holm project was out of the reckoning, threw up a potential deficit of some £1,700.

Hunt gently reminded the Trust that his remuneration was shockingly low: he had received only £750 in fees for the year. That was exactly half the amount paid to Alderson, and both were effectively putting in many unpaid hours on behalf of the Trust.

In September 1974 Alderson presented his first annual report. On behalf of the Trust, he had travelled 15,597 miles in his first year in the post; he had made advisory visits to 38 members, carried out the breed survey, attended agricultural shows, set up and implemented major projects, attended committee meetings and so on. He calculated that, in real terms, his net return from all his work was about £5 a day and that there was plenty more work he felt he should be doing.

The September meeting also noted that the Trust was having minor rebellions among one or two breed societies. Back in 1971, at the first joint Working Party meeting at the NAC, the Longhorn Cattle Society had expressed its concern at its splendid breed being seen merely as park decorators rather than the commercial animals it was felt they could be (as would be proved). Now the Trust had a plaintive letter from the Southdown Sheep Society, a venerable institution which objected to its once widely influential breed being included in the Trust's 'second interest' group — very demeaning to a breed which, with its origins dating back to the famous 18th century breeder John Ellman, had spread worldwide and had spawned several other downland breeds. There would be other breed societies who were similarly offended over the years.

There were so many philosophies to be sorted out and projects to be considered during those first two years of the Trust. Everything was happening so fast: there was a great sense of urgency and perhaps a tendency to be too enthusiastic. At its second field day and conference in October 1974, held at Leigh Park in Hampshire, there was a slightly tongue-in-cheek but nonetheless percipient talk by the Devon veterinarian, Jim Hindson, which came to be known as the Three Black Spots Syndrome paper.

Reading the early issues of *The Ark*, one can sense that excess of enthusiasm in the pursuit of rare breeds that sometimes led people up intriguing but perhaps irrelevant garden paths — but that is with hindsight. At the time there was a strong zeal, a great sense of purpose and dedication to a cause that people found exciting. The concepts were new, the buzz from the reactions generated by publicity was high, and, in spite of differences of opinion, there was a tremendous sense of being in a friendly club of people with a common interest in the breeds, regardless of individual backgrounds or motives.

The early strands all persisted — zoologists, pedigree breeders, farm parks, researchers, academics .... and increasingly individuals with an emotional interest in what they liked to call a 'living heritage'. Smallholders also began to join the membership, recognising that many of the rare breeds were ideal for non-commercial, low-budget holdings: the pigs were still basically cottagers' pigs, the sheep had interesting fleeces for hand-spinners, some of the cattle were ideal milch cows and sucklers, and so on. Perhaps the geneticists began to shudder, just a little.

SADDLEBACK PIG

SETTLING DOWN

At its last 1974 meeting, in November, the Advisory Committee reviewed the remuneration of its two paid consultants, Hunt and Alderson. It was very mindful of the debt it owed them both. In the case of Alderson, they were aware that his present remuneration did not cover all his expenses but they expressed mild concern that he was carrying out other consultancy work and perhaps visiting too many agricultural shows, and they decided not to increase his basic remuneration but to look for ways and means of providing specific payments for projects and other special work, an arrangement which certainly disappointed Alderson. They increased Hunt's remuneration to £1,250 per annum.

They then agreed a programme of events for the coming year. Among them would be the first Rare Breeds Show & Sale, which would eventually become a major calendar date for the Trust and its members. Another important development in 1975 was the formation of the Trust as a limited company, in line with the original concept, and the consequent development of a committee structure under a council of directors. Peter Hunt remained deeply involved, a little to his surprise: he had not really anticipated continuing after he had reached his goal of setting up the company. By March 1975 the membership had doubled to 917, and Hunt's home was still (along with Alderson's) the Trust's office.

The Trust continued to consolidate its activities: an AI programme and semen bank, approval of rare breeds centres, the Combined Flock Book (still a private rather than Trust concern), Linga Holm and Lower Midgarth. There were several international developments as well. The AI project and semen bank proposals by Alderson were accepted in January; he also set up a small technical advisory group during the year as a sounding board for new ideas and developments and was spreading the Trust's message by means of talks and articles.

Others were also very active on the public relations front. Hunt, in August 1975, was in contact with various show societies and told them all about the Trust: that its membership was over 1,200, that they had carried out a breed survey in 1974, that they had successfully launched the Linga Holm project, the AI programme and semen bank, and would shortly be holding their first show and sale for rare breeds. And Joe Henson seemed to be everywhere, talking about his farm park and about rare breeds in general and the Trust's work in particular.

On the popular front, sculptor John Harper had designed collection boxes based on a two-foot-high model White Park bull. Hunt well remembers having a herd of them spilling out of his garage at Kings Langley, but they were easier to handle than the stuffed bull's head he had been carrying around to shows.

There were open days — at Rippon Hall, for example, which attracted more than 200 — and field days at the newly recognised Hele Farm Park rare breeds centre near North Bray and Dartmoor, owned by Hedley Le Bas. (It should be noted that, despite the similarity in names, Hele Farm Park is not the currently well known East Hele Farm owned by Anne Petch.) Another was held at Michael Ann's Drusillas Zoo Park in Sussex.

A Limited Company

In March 1975 the Advisory Committee held its final meeting and formally agreed the transition to a company limited by guarantee, without share capital, a status which was achieved in May 1975 and persists even now. The first meeting of the directors who formed the new company's Council took place on 16th June. The company formally commenced to trade on 1 July 1975, when it took over all the existing assets and liabilities of the old Trust. At that stage the assets totalled a healthy £21,240 including some £6,000 in property (Linga Holm) and display equipment, library and photographs; and £1,614 as the value of the North Ronaldsay sheep at cost.

The second Council meeting, in August 1975, considered a range of subjects that reflected the expanding interests of the Trust: the forthcoming first Show & Sale; grant aid for the Lower Midgarth field centre project; a revised priority list on the basis of the 1974 survey; and an agreed breeding programme for rare breed pigs, including the use of AI and a semen bank and the possible importation of fresh boar stock from Australia, which was quite a drastic effort to bring new life into the seriously endangered pig breeds. By now the Trust's membership had increased to about 1,200.

Little breed controversies continued. At a Council meeting in October 1975, for example, there was a report on the Blue Albion. In November it was the Leopard Spotted Horse that stirred up a bit of argument and something of a personality clash within the Trust.

Little niggles over the remuneration of Hunt and Alderson persisted. Both men were giving the use of their own offices as well as their own time and expertise, and Alderson estimated his net return in 1975 as all of 87p per day.

The company's first annual general meeting and the first of a series of Scottish symposia were held in December 1975.

Setting the pattern

By now the Trust was getting into its stride and sorting out its procedures, setting a pattern for the future. The first three years had seen many new schemes put in place and philosophies clarified; in 1976 these were consolidated. Acceptance procedures were laid down, priority lists for the breeds were established, a more formal programme of breeders' workshop was devised and a register of genetic defects was quietly compiled. The boar importation scheme became a reality, the Trust membership continued to rise steadily, and the officers continued to be underpaid.

In January 1976 the newly elected Council held its first meeting. Chapter Three looks at how various committee structures began to develop, and at the people who became chairmen, presidents, treasurers and so on, as well as the growth of the membership as time went by. Chapter Four looks at the early projects in detail while other innovations in this period are described in relevant sections elsewhere. 1976 proved to be a very creative year.

A pattern of events was taking shape: breeders' workshops, a second Scottish symposium and the second Show & Sale at Stoneleigh. There were more fund-raising ventures, including liaison with the Joint Charity Card Association. Gradually the aims and ways of achieving them were growing clearer with experience and Alderson outlined a programme of projects based on the three stages of conservation, evaluation and utilisation. This was a considerable step forward: until then the Trust had been largely concerned with rescuing the breeds from extinction, without too much thought about what would happen to them after that. The sense of merely reacting to crises was being replaced by a more constructive and proactive attitude to rare breeds.

CHAPTER THREE
GROUNDWORK

NORFOLK HORN RAM

BERKSHIRE GILT

CHAPTER 3:
GROUNDWORK

This chapter examines the development of the Trust's various committees, its growing membership, its offices and officers, and its funding. It is about the infrastructure which enabled the work of conserving rare breeds to be carried out sensibly and effectively. First of all, however, it looks at something even more fundamental: the major statements of Trust policy upon which everything else is based.

POLICY STATEMENTS

The Trust has always been an organic creature: in the early years it evolved gradually and its philosophies and policies have developed almost by stealth. A group of people, or more usually an individual, created a policy which seemed to make sense at the time and became a precedent, sometimes without any real analysis. It was just absorbed and quietly adjusted until in due course it had become part of the Trust's policy. Sometimes there was no formal recognition, but there was acceptance.

At intervals, however, people would stop and think about a major theme and look at it deliberately, either as individuals producing papers for the Trust to consider, or in the occasional think-tank sessions that became landmarks during the Trust's history — starting with that 1971 Task Conference which had given the final impetus to form the national organisation in the first place.

The first deliberate step in policy-making was in March 1969, when the old Working Party defined the meaning of the word that was the essence of the movement. They recommended 'a definition of "rare", which essentially means where the breed is reduced to five tups and twenty ewes in an active breeding condition and/or in danger of being entirely used for crossbreeding'. (At the time, the Whipsnade gene bank largely consisted of sheep, and their main concern was the Norfolk Horn.) Technically, such a population size was hardly effective — possibly enough for survival but not for genuine conservation — but it was a start.

Apart from its own Memorandum and Articles of Association on the formation of the company in 1975, the Trust's first formulated policy was the procedure by which a breed is accepted, i.e. one that falls within the Trust's care and receives its recognition and support. This procedure, which is fundamental to all the Trust's activities, was first drawn up in its proper form in September 1976 and is essentially the same today.

On 9 October 1979 Alderson wrote a 'Definition of Trust Policy' which attempted to stand back and look at what the Trust was all about. First of all, his paper made it quite clear that the sole purpose of the Trust was to conserve animal genetic resources. Then he looked at how this might be done, and what the Trust was already doing. The full text of the document is given here because it tells a great deal about the Trust's aims and the Technical Consultant's views on how those could best be achieved.

Definition of Trust Policy

1. What is the purpose of the Trust?

The purpose of the Trust is to conserve animal genetic resources. Although the Trust has been involved to some degree with supporting genetic conservation in other countries, its main concern is with the genetic resources contained in British rare breeds of large livestock (i.e. cattle, sheep, goats, pigs and horses). The breeds which qualify as 'rare' are identified by the Rare Breeds Acceptance Procedure.

2. How can the purpose of the Trust be realised?

The conservation of genetic resources will be achieved either by direct action with the relevant breeds of livestock or by indirect action through the breeders and owners of these animals. Some benefit may be derived from other interested parties on a purely financial basis. The most important methods of realising the purpose of the Trust can be categorised as follows:

- (a) Preventing rare breeds becoming extinct.
- (b) Long-term storage of frozen embryos and semen.
- (c) Creation of new breeding units:
    - i. independent units
    - ii. units within larger commercial units.
- (d) Increasing the numerical status of each rare breed.
- (e) Improved management of rare breeds.
- (f) Evaluation of rare breeds.
- (g) Promotion and utilisation of rare breeds.

3. How can these methods be implemented most effectively?

(a) Although the Trust has compiled a list (1974) of breeds of livestock that are in danger and merit support, the mere knowledge of the numerical status of a breed does not guarantee its survival. It is necessary to maintain a continuing, up-dated *information* bank. The necessary information will be obtained on a regular basis by direct contact with breeders and Breed Societies, supplemented at intervals by a detailed *Survey* of endangered and potentially endangered breeds and by specific research and trials.

(b) As a long-term policy and as an insurance against the disappearance of all the living representatives of a breed, the storage of *frozen semen* for cattle breeds at this stage, and the storage of *frozen embryos* for all species, when the necessary technological advances have been made, are of great importance. The *Bank of Genetic Variability* already exists, but should be extended to include as wide a selection of bulls as possible from each generation of each breed. The selection of the bulls (and embryo parents) should be based on their pedigree and an *inspection*. They should be a representative sample of the total population and they should be of sound constitution and free from defects.

(c) The creation of *significant breeding units* of rare breeds received high priority during the early years of the Trust's life. It remains an item that should receive a good deal of attention, but now greater emphasis should be placed on persuading owners of *large commercial units* to include a small group of a rare breed within such units. This is likely to give several benefits; the standard of management can be expected to be above average; the rare breed can be evaluated alongside popular breeds; and in most cases accurate and detailed *records* will be available to feed into the information bank. The security of a breed increases in proportion to the number of breeding units, so that even small units can play an important role.

The *procurement of livestock* is a necessary part of establishing a new unit and in this context the *breed organisations*, the *Show and Sale* and *The Ark*, have a valuable role to play. They are able also to provide focal points to facilitate the *exchange of breeding stock* between existing breeders. The purchase of livestock must be based on a planned programme which is designed to control inbreeding. Breeders should be advised regarding *breeding programmes* which can take advantage of contract matings, AI, etc., and which should actively maintain existing bloodlines. The *structure of each breed* should be analysed and weak lines given *extra incentives.*

(d) The increase in the numerical status of the rare breeds should result automatically from the Trust's activities. The natural increase in numbers, that derives from the creation of new breeding units and the increasing confidence and involvement of established breeders, provides the most secure and long-lasting foundation for the future viability of a breed. In some cases extra stimulation may be necessary. This can be achieved either by intensifying the tempo of the Trust's activities, or by special projects, such as an *embryo transfer* programme. In any breed where numbers are increasing significantly, a close watch must be maintained on the quality of the animals with special regard to *congenital defects*, which should be recorded and eliminated. To achieve this the necessary information must be collated and analysed to provide the results on the basis of which decisions can be taken.

(e) The involvement of a disproportionately high number of breeders who are relatively inexperienced in livestock husbandry, and the special characteristics of at least some of the rare breeds of livestock, makes the provision of *advice on the management of livestock* a matter of high priority. It is of little use saving a breed and devising carefully formulated breeding programmes if the management of the animals inhibits them from growing, reproducing and performing successfully.

Advice can be made available to breeders in various ways. At the most elementary level organisations such as the ATB provide *courses for beginners*, although these are directed to breeders of commercial livestock, and this applies also to a series of *ADAS leaflets*. However, much of the information provided is relevant. More specific advice can be provided through the Ark or by *letter and telephone*. The amount of advice directed to individual breeders will be limited, although it could be justified in the case of the *larger rare breed units* and *Approved Centres*. For other breeders a programme of *breeders' workshops, technical meetings* and *seminars* should be available to give advice on a group basis.

(f) The evaluation of rare breeds, and the definition of their qualities and characteristics, is an activity of secondary importance, but nevertheless it forms an integral part of the long-term survival of these breeds. A considerable element of evaluation will arise naturally from the records maintained by breeders. These records should be collated and stored in *the information bank*. In addition specific qualities can be measured as part of a purpose-designed project e.g. *Ease of Parturition, Bull Performance Test*, and added to the information bank.

(g) The promotion of rare breeds is the final essential ingredient of a total package that is calculated to ensure their survival. The dramatic and beneficial effect of publicity on the fortunes of various breeds has been demonstrated already. The annual *Show and Sale* and *The Ark* both play an important role in this progress. There is a need to achieve wider coverage in those *magazines and journals* which reach breeders and potential breeders who are likely to play an active and constructive role in maintaining breeding units of rare breeds.

4. Trust policy and its implementation

Taking the purpose of the Trust as defined in item one as the starting point, the Trust has been successful to a large extent in achieving its initial objectives but has been deficient in consolidating its position, partly due to limited financial resources. This limitation still obtains and there is a need to identify those aspects of the Trust's work which should receive priority.

From the details included in items two and three above, it is clear that only a minority of the membership of the Trust is actively concerned with conservation of rare breeds. These members should be the recipients of the resources and services provided by the Trust. The priority members (i.e. those who are contributing directly to genetic conservation) need to be provided with a more comprehensive range of services than they receive at present in order to help them to continue and improve the work that they are doing. If these services and direct conservation measurers cannot be implemented the reason for the Trust's existence ceases:

(a) Conservation policy and strategy. The basic philosophy of conservation should be kept under regular review and the Rare Breeds Acceptance Procedure should be applied in the revision of the list of 'priority breeds'.

(b) Maintenance of an information bank. This will contain all information relating to rare breeds and genetic conservation. Data should be obtained from all possible sources, e.g. Breed Societies, breeding units, projects, etc., and the analysed information should be used to provide an up-to-date appreciation of the position of each breed and to enable an effective advisory service to be maintained.

(c) Advice on the breeding and management of rare breeds. This will include direct advice to larger breeders, Approved Centres, and to groups of smaller breeders. The breeders' workshop programme, reinforced by other technical meetings and seminars, and by correspondence and direct telephone contact, can provide a wide range of management advice. Advice on breeding will be provided most effectively either directly to larger breeders or through the medium of group programmes.

(d) Creation of new breeding units. New units, either as a separate entity or within a larger unit, involves direct contact with potential breeders and the provision of advice regarding breed selection, procurement of breeding stock, breeding programmes, management system, recording, etc.

(e) Inspection of livestock. With the necessary development of the semen bank and contract mating, and the possible introduction of embryo transfer, embryo storage and pig AI, there will be an increasing need to select the breeding animals for each purpose.

Inspection of animals for congenital defects would be desirable.

(f) Show and Sale as a focal centre for rare breeds promotion and the procurement of breeding stock.

(g) The Ark as a medium for rare breeds promotion, for the procurement of breeding stock, and for providing advice to breeders.

(h) Projects. The following projects in the Trust's current programme deserve high priority: Survey, Breed Incentives, Breed Structure Analysis, and Register of Congenital Defects.

In summary, the main part of the Trust's resources should be directed more effectively towards the realisation of its primary objective, namely genetic conservation, and to those members who are contributing positively to this objective.

Alderson was drumming the same drum in 1991 when he drew up another major policy document in his new role as Executive Director of the Trust. At the time there were mutterings from some of the Council members to the effect that policy was the responsibility of Council and that its executive staff should simply carry out such policies. Others, including the Trust's president Lord Barber, believed that councils should appoint the calibre of executive staff who could be trusted to get on and do things without constant reference to committees.

Alderson found himself in a peculiar situation. He had worked actively in the Trust for longer than anybody else who was at that stage on Council; he had over the years conceived and defined Trust policy. Yet while there were those who appreciated his continuity and long experience, one or two wished to restrict his influence. He therefore produced a draft policy document for consideration by Council and, in the event, the document was essentially accepted as it stood and became official policy. A small working party was set up to look in more detail at the Trust's procedural standards, particularly with regard to approved centres, support groups, regional sales and so on.

Alderson's 1991 document, like that of 1979, sought to encapsulate the whole concept of Trust policy. It also established the principle of a new structure whereby Council was responsible for taking major policy decisions and establishing the broad policy of the Trust, but that the two managing committees (Executive, and Breed Liaison) and the executive staff were responsible for carrying out that policy.

One of the planks of Alderson's document was that the chairman should drill into his Council that they were charity Trustees and that they should not take on that role lightly. Legally, they can be held responsible for any deficits or liabilities which might arise.

COUNCILS AND COMMITTEES

There are those who love committee work — the power games and the debates, and (on the best of them) the productive exchange of creative ideas. And there are those who wouldn't go on a committee if you paid them. Some change their minds: Geoffrey Cloke is a good example of a man who had never been near a committee in his life, but once he found himself drawn into the web of the Trust's committees he discovered that it was possible to become involved in the creative debate while remaining detached from the personality clashes that arise when any group gets together to discuss matters about which individuals feel strongly. Within any organisation, there is inevitably a tendency to build little personal kingdoms, or cliques, and that, perhaps, is all part of belonging to an organisation. The challenge is to use those human characteristics constructively to the good of the organisation as a whole.

In an attempt to clarify the setting, here is a Dickensian checklist of names — not of the cast of characters, but of groups. In the early history of the Trust so far, there has already been mention of the ZSL's *Breeding Policy Committee* (BPC) and its *Gene Bank subcommittee* (GBSC). Next came the exploratory *joint meeting* in 1967 between ZSL, RASE and other interested parties, which began as an informal group but, once the NAC and Reading had accepted the gene bank sheep, Rowlands and Dadd saw the need for a technical panel to give guidance on the animals' management. The idea had crystallised at a meeting at Stoneleigh, and what emerged was the Rare Breeds of Domestic Livestock Gene Bank *Working Party*, which held its first meeting in March 1969, not only to look at the management of the gene bank sheep but also to take a much wider look at the future for rare breeds and with the specific task of setting up the Trust. This cumbersomely named group was the real germ of the RBST: with the support of the Task Conference in 1971, it decided the best way forward was to create a national organisation to care for the interests of rare breeds.

In due course the Working Party decided that the new organisation should be a trust and it became a *Steering Group*. Its change of name from Working Party to Steering Group was as late as February, 1973, just a month before the Trust became formally recognised as a charity.

In May 1973 the three Trustees from the old Working Party appointed an *Advisory Committee* (AC). This was the Trust's first committee and was also intended to be the basis of what would be a Council once the Trust became a company in 1975. The candidates for this committee and future council had been established at a Working Party meeting in April, 1972. The AC held its final meeting in March, 1975, when it formally approved the transition to a limited company, and the first meeting of the company's directors — its *Council* — took place in June 1975. Half its members were (and still are) nominated by corporate members and half by individual members, though others are co-opted from time to time. Council members stand for a three-year term. Initially they were the members of the old Advisory Committee which automatically became the Council when the company was established. Then in December 1975 came the first elections, and the first meeting of the *elected* Council was held in January 1976.

A structure of committees, subcommittees and advisory groups to look after various areas of interest began to evolve in the early days of the Trust, especially when Michael Rosenberg was elected to Council. Before, everything had been a little happy-go-lucky but he, with a dominant personality, an analytical mind, a shrewd business brain and a natural affinity for creating systems and procedures, soon began to encourage a more organised approach, with the help of Peter Hunt in his administrative role.

The first was a *Technical Advisory Group* (TAG) formed in 1977. Its members were initially Christopher Dadd, Jim Hindson, Idwal Rowlands, John Bowman and Peter Jewell — all of them (except Hindson) involved back in the Working Party days. By May TAG was discussing, for example, ova transplants in sheep and group breeding programmes relating to a specific rotational mating scheme devised by Alderson in 1974, for which it was hoped that the Trust might buy females of selected breeds and place them on members' farms on a co-operative basis so that the animals could be included in the programme.

The group had come together because Alderson needed a sounding-board. As he had written in December 1976, he wanted them to 'comment on technical proposals before they are presented to Council so that your expertise is fully exploited even if you are unable to attend Council meetings.' He explained his reasons for forming the group:

> Several times recently I have felt that decisions have been taken without a real understanding of the basic philosophy of genetic conservation and I think that the first task of the Technical Group should be to define the objectives of the Trust in a language that the members of Council can

understand. There is no longer a problem in determining which breeds should be included, as we now have the agreed Rare Breeds Acceptance Procedure. The main problem seems to be caused by the incompatible requirements of breed improvement and the maintenance of genetic variability and this potentially draws the Trust into conflict with breed societies and commercial breeders. If you agree with this viewpoint, we must present it as a proposal of basic principles which will act as a foundation on which to build policy decisions and long-term programmes.

The TAG would become absorbed into other committees in due course: by 1982 it was the *Technical Advisory* committee, still chaired by Jim Hindson and still with Jewell and Bowman, though Rowlands had by then retired; new members included Juliet Clutton-Brock of the Natural History Museum, Ann Wheatley-Hubbard from the RASE, George Jackson of the NAC and Geoffrey Cloke. Before then Alderson had already set up a five-year project programme with TAG's help and with essential financial backing from Rosenberg. The projects included surveys of breed populations, blood-typing and milk-typing, analysis of breed structure, ease of parturition studies, and breed incentive programmes.

In September 1978 Alderson wrote to his TAG inviting them all to attend a meeting of the *Standing subcommittee* (SSC) in order to discuss procedures for dealing with major items which were of common interest to both groups. This body had already changed its name once or twice: it would end up as the *Executive committee* but at first it was known simply as 'the subcommittee' and then the Chairman's subcommittee before formally becoming the SSC in January 1976. In its original form it was to consist of the chairman, vice chairman, treasurer and two other Council members of the chairman's choice. Its members in the early stages were Joe Henson, Michael Rosenberg, Christopher Dadd, Denis Vernon and Richard Cooper, soon to be joined by Geoffrey Cloke and John Hawtin.

The *Council* itself at this stage was somewhat passive, relying naturally on its two officers (Peter Hunt as Secretary and Lawrence Alderson as Technical Consultant) for many initiatives. That was, after all, their job on a day-to-day basis as the Trust's paid staff. Council members also had positive ideas: for example, Rosenberg was the father of the Show Demonstration programme, and Jim Hindson was the first to suggest that a register of genetic defects should be established. It is clear from the minutes of the period that the SSC was taking many of the Trust's policy decisions, with the Council to some extent acting as a rubber stamp. Over the years the strength of the Council's role would vary, according to the personalities involved, but there is a good case to be made that it should not become bogged down in the everyday detail of the Trust's activities.

There was a gradual turnover in the membership of the Council. For example, in 1976 Michael Ann (one of the original farm park owners) dropped off, and the following year John Bowman decided that his busy schedule prevented him from attending meetings, while John Taylor did not stand again due to ill health. Taylor was an interbreed cattle judge at the first Show & Sale.

A comprehensive table of Council membership over the years is given in the Appendix but some members are of particular interest. In 1987, Sir Derek Barber agreed to be co-opted to Council and in due course, as Lord Barber of Tewkesbury, he became the Trust's president and a most effective and imaginative participator in Trust affairs. His experience in other organisations such as the Countryside Commission, the Royal Society for the Protection of Birds and the Royal Agricultural Society of England (of which he served a term as president) was invaluable.

At its meetings in the autumn of 1976 the Council took a close look at committee structure and considered a proposal that the business of Council should be divided and in part delegated to two committees: a Scientific Projects committee and a Breeders and Membership committee. New names came into the picture: among those suggested as members of the latter group was Christopher Marler, the innovative owner of a collection of waterfowl and wild animals and a respected cattle breeder. However, this somewhat arbitrary division into two committees was rejected.

In August 1979 another new committee structure was agreed. There would be essentially four groups under Council: the chairman's (the *SSC*) and technical (*TAG*), and two new ones: a *Breed Promotion and Liaison* committee and a *Fund-raising and Promotion* committee.

By 1982 the *Breed Liaison* committee, chaired by Geoffrey Cloke, had become an important group and included, for example, Ken Briggs as chairman of the Linga Holm working party, Arthur Manchester (Secretary of NPBA), Clem Pointer (Longrigg's successor as chief livestock adviser, ADAS), and Ann Wheatley-Hubbard. In 1990 the BLC was 'rationalised' into subcommittees: it remained the main group but there were also subcommittees for different species — pigs, horses and cattle, sheep and goats.

The *Fund-raising* committee was chaired by Richard Cooper (treasurer in 1975—7), who had many useful contacts in this connection, as did members Christopher Dadd and Christopher Marler. Other members were Michael Rosenberg and Denis Vernon (treasurer from 1978). The committee ceased to exist at the end of 1985. The whole subject of fund-raising is considered separately later in this chapter.

Two other groups had evolved by 1982. The first was the *Linga Holm* subcommittee, reporting to the Breed Liaison committee and chaired by Ken Briggs; its members included Peter Jewell and also three veterinarians — Jim Hindson, Bill Carstairs (who still practised in the Orkneys) and Marshall Watson.

The second was *The Ark* subcommittee, including Rosenberg and Alderson as the journal's founders and Judy Urquhart, assistant editor to Alderson at that stage. It was in 1980 that Rosenberg and Alderson finally handed the journal over to the Trust. At the same time they handed over the Combined Flock Book, and a *Combined Flock Book* panel was formed under the chairmanship of Ken Briggs as a BLC subcommittee; later, under the chairmanship of Dr Richard Harper-Smith, it became part of the new *Sheep and Goats* subcommittee.

By 1986 the committee structure needed another good shake-up and John Wood-Roberts (Secretary, 1983—8) gave it exactly that, making the whole system more rational. The Technical Advisory committee had been divided into two — a *Project Development* committee, and a *Scientific Advisory Panel* under the chairmanship of George Jackson — but it was not a happy arrangement: the SAP members felt that they had little impact and that most decisions were taken by the new Project Development committee, and so SAP was disbanded and its members were absorbed into the Project Development committee, chaired by Robin Mulholland, chief ruminant officer of ADAS, who later transferred to the Crown Estates. Mulholland, who had been an outstanding schoolboy swimmer, shared with Dudley Reeves a passion for angling.

A *Support Groups* subcommittee had been formed and was responsible to the SSC, which was renamed the Executive committee. There was a subcommittee to look after *approved centres and farm parks*; and there was a *Show and Sale* subcommittee which operated under a series of show directors — Alderson (1975 and 1976 shows), Rosenberg (1977—81) and John Hawtin, who remained chairman of the Show & Sale subcommittee until 1993, when other members included Frank Bailey, Jonathan Cloke and Alan Lyons. The group was directly responsible to the Executive committee.

To summarise, in 1986 the new structure was based on a *Council* to which three main groups reported: the Executive committee, Breed Liaison committee and Project Development committee. Responsible to the Executive committee were The Ark panel, the Support Groups, the Trading Company and the Show & Sale committee. Responsible to the Breed Liaison committee were the Linga Holm committee and the Combined Flock Book's Registration and Inspection subcommittee.

In 1987 it was agreed to amend the 34th paragraph of the Memorandum & Articles of Association to read: 'The Council shall determine from time to time the formation of committees and shall appoint the chairmen thereof. Such chairmen, excluding those of subcommittees and panels, shall be ex officio members of Council.' Clearly Council thought it was time to take more control over its own committees, and a few years later it also tried to control the membership of each committee. But that was all part of the see-sawing balance of power between Council, the Executive committee and perhaps various individuals within the Trust, including chairmen.

CHAIRMEN

The Trust has had seven chairmen since its foundation in 1973, each with very different personalities, different styles of chairmanship and different influences on the way the Trust operated.

Joe Henson

The first chairman was Joe Henson, a man with a self-confessed liking for showmanship and a flair for publicity. He was also a shrewd businessman on behalf of his Cotswold Farm Park: essentially Henson believed that whatever was good for the latter was good for rare breeds in general and thus to the benefit of the Trust. He had an infectious enthusiasm which served the Trust well in gaining new members.

He had originally been invited to join the old Working Party in 1969 because of that enthusiasm and as a farmer with an active and practical interest in rare breeds. He became one of the three initial Trustees and was appointed as the first chairman of the Advisory Committee of the new Trust (1973—5) and subsequently as chairman of the Council (1975—7) when the Trust became a company.

His second term of office (1986—8) coincided with a period of jostling for position with the executive staff and he took the post with mixed feelings: he thought that it was almost a retrograde step and that the Trust should be taken over by younger members instead of the faithful old guard.

Henson remains a firm believer in a 'grass roots' movement with a broad membership base balanced by adequate technical expertise to counteract any tendency among breeders to take the 'fanciers' road to show-ring aesthetics. He was invited to become a vice president of the Trust, which he considers 'the greatest honour I have ever had'. Although he is no longer on the Council, his daughter Libby Henson is a member of it and actively carries on the family involvement in Trust affairs.

Ann Wheatley-Hubbard

Henson is an informal man and that was reflected in the way he chaired Council meetings: he was more than happy for people to talk themselves out if they wished. His temporary replacement in the chair in 1976 had a marked contrast in style. Ann Wheatley-Hubbard proved to be incisive, firm, very efficient in running Council meetings and thoroughly down-to-earth. She had the art of ensuring that meetings were brief and to the point but in such a way that all those present felt satisfied that their views had been properly aired. She knew how to let people have their say but without allowing them to 'rabbit on', and she was greatly respected for it. She had a solid agricultural background: her grandfather had started the family herd of Tamworth pigs in 1922; her father continued the herd at Berkswell, a few miles from Coventry. When he died suddenly in 1943, she took over the running of the estate at the age of

19. Ten years after her marriage she took the Tamworths to a large estate at Boyton Wiltshire, where they are now managed by her son and daughter-in-law.

Ann Wheatley-Hubbard is the sort of woman who simply gets on with it. Her father had been president of the NPBA and a member of the RASE Council in his time; she herself became the first woman member on that council in the 1950s and has been a member ever since. She became a Trustee of the RASE in 1960 and has been president of the NPBA in the true family tradition. She represented the RASE on the Working Party that had been formed to create the rare breeds organisation and became its vice chairman. Having served as acting chairman of the Trust's Council in 1976, she became its chairman again from 1981 to 1983. She is now one of its vice presidents.

Richard Cooper

Richard Cooper (now Sir Richard Cooper, Bt) was a very different chairman. He served first from 1978 to 1980, when he was farming cattle and corn in Dorset, and again in 1990—92. He also chaired various committees in between. He first became involved with the Trust through his association with the RASE, of which he was a council member. Those who know him well would probably agree that he had no great liking for the hurly-burly of controversy and confrontation within the Council: with his easy charm and polite manners, he would feel much more at home in the clubs of St. James's — or even on the steppes of Kazakhstan, with an advisory brief! He had the ability to open doors gracefully and, with many good contacts, he brought useful sponsors and donations into the Trust right from the start. He was a cajoler by nature and also full of ideas, some of them less practical than others but at least he generated them. He held an annual Smithfield dinner which was a very good venue for bringing new potential contributors into contact with the work of the Trust.

An example of Cooper's way of working can be given by some correspondence in January 1978. Cooper had made good use of his network: he wrote to many important people informing them of his chairmanship as a way of introducing the work of the Trust and suggesting that they or their organisations might like to help the Trust in some way. Lord Nugent of Guildford, VC, responded as chairman of the National Water Council, first of all by saying: 'Congratulations, and perhaps I should add commiserations ...,' which was perhaps indicative of the general opinion of the Trust in those days. Cooper, however, had clearly been imaginative in this approach: he realised that regional water authorities had land around their reservoirs which might usefully be grazed by some of the primitive sheep breeds, for example. One day this idea would lead to Soays being used to good effect on the waste tips of English China Clay.

Alderson recalls that, under Cooper's chairmanship, there seemed to be a much greater two-way flow of information in advance of Council meetings. Cooper wanted to be a 'hands-on' chairman, actively involved in the work of the Trust. He liked to attend the breeders' workshops, and kept his own flock of Leicester Longwool sheep.

Geoffrey Cloke

Geoffrey Cloke was chairman from 1983 to 1986 — he was made Executive Chairman after the rather sudden departure of Michael Rosenberg. Cloke has always been a pig-and-poultry farmer; he came from several generations of farmers, originally in Devon and then Gloucestershire, where his grandfather had kept Gloucester cattle. A great grandfather had a flock of Oxford Down sheep in the late 19th century; other members of the family bred Shorthorn cattle, and in the late 1930s Cloke remembers his father receiving the family's first Large Black pigs ('three little black things, in a box'). Since then, Cloke has kept several rare breeds of pig, especially British Lop, apart from his commercial herd of free-range Welsh. He was the first individual (as opposed to company) to bring the Duroc into the UK, in 1973; he kept them for three years but nobody else seemed to want them in those days because of their thick skins and the leep hair follicles which meant that they did not dress tidily. He then looked for a 'white Duroc' and came up with another American breed, the Chester White, which he imported by means of embryo transfer in 1979.

For as long as he can remember, there had always been a flock of black hens around the place: it was not until much later that he realised that these Croad Langshans had become a rare breed. He still breeds them, and White Sussex, but his most serious passion is for pigs, always on extensive systems. He joined the NPBA in 1945 and became its chairman in 1984; after a very long stint in that position he became president of what is now the British Pig Association in 1993. He also took on a three-year term as president of the European Federation of Pig Breeders.

Cloke was approached at the time of the 1970 Reading University students' survey of rare breeds and has been involved with the Trust directly or indirectly ever since. He went to the Task conference in 1971 and joined the Trust as a founder member in 1973; he was asked to join its Council in 1976 for his expertise in pig breeding (he is still a Council member) and was its vice chairman from 1981 to 1983 and from 1992 to 1994 (as chairman elect). He became chairman of the Breed Liaison committee at the end of 1980 and at the same time a member of the SSC.

As chairman, Cloke managed to combine light-heartedness with effectiveness. His meetings were relaxed and good-humoured but they also dealt efficiently with the business in hand. His general attitude to the Trust's work highlights the way in which the rare breeds movement can fire people with enthusiasm for its cause and compel them to work for it with little or no financial reward. Even its two earliest paid consultants, Hunt and Alderson, worked more for love than money at very low fees in comparison with the hours they contributed — and they were, after all, in the insecure position of being self-employed. All the Council and committee members act on a voluntary basis, and all those who help on panels, or at the Show & Sale, or on support groups. This huge well of voluntary effort has been a significant feature of the Trust since its inception.

Michael Rosenberg

At the heart of this well was the man whose financial contributions have been crucial for the Trust's own survival, Michael Rosenberg. He was far from content simply to sign cheques: he worked, very hard, for the Trust and put in countless hours and endless energy stumping round the countryside on its behalf, taking its message to the shows, shaking its organisation into shape. This colourful American earned, many times over, his reward of an Honorary CBE in 1986 for (in the words of the citation) his 'unique contribution to genetic conservation through the survival of rare breeds'. In making this commendation the Minister of Agriculture (Michael Jopling) went on to say that Rosenberg's 'unfailing energy and enthusiasm' as Honorary Director of the Trust had made the latter an 'important preservation organisation'.

Rosenberg was chairman of the Trust from 1980 to 1982 and in 1985 he accepted the office of vice president, though within months he deferred his acceptance. Meanwhile he had been its honorary director but resigned from that post at the end of 1985, a year in which he had also become an honorary vice president of the RASE.

Denis Vernon

Another strict Council chairman was Denis Vernon (1988—90). Vernon is a Geordie with legal qualifications; he has achieved success in the business world and lends his support to several charities, especially since his retirement from being chairman of Ferguson International Holdings PLC in 1991. He joined the Trust in 1973 and has served as its honorary treasurer since 1978.

As chairman, Vernon was not averse to cutting off a rambling discussion, often quite brusquely: his approach was businesslike which, unfortunately, rankled with some of his Council members and generated a minor 'palace revolution'. As in so many small groups, there was a constant if low-key power struggle and the balance swung between the Council, its executive committee and its executive staff.

Dudley Reeves

Dudley Reeves, who succeeded Ann Wheatley-Hubbard as the RASE's representative on the Trust's Council in 1984, became chairman in 1992 and found himself with something of a handful: a fairly vociferous Council with several fresh faces who brought it well and truly to life. Reeves's father, who established a charitable trust from which the RBST benefited, farmed Red Poll cattle and Tamworth pigs. Reeves himself, an RASE Council member, retired from farming in Buckinghamshire recently but maintains his strong interest in conservation; he has, for example, chaired the management committee of FWAG (Farming and Wildlife Advisory Groups).

The most open and approachable of chairmen, Dudley Reeves was interested, committed and helpful. He effected gradual but essential change by bringing on the next generation of potential chairmen and invited Alan Black and Anne Petch to sit on the Executive committee.

A MEMBERSHIP ORGANISATION?

There is a fundamental philosophy which has often been debated within the Trust, especially over the last decade or so. Should it be simply an organisation devoted to genetic conservation, essentially a small group of people actively breeding the livestock? Or should it be a large and broadly based membership organisation? There are arguments on both sides.

Alderson has always argued that the Trust should concentrate on those people who actually keep rare breeds, and that the Trust's funds should be devoted directly to conserving and promoting those breeds.

Rosenberg, among others, agreed with this view but it led him to a slightly different philosophical approach. He felt that the breeders keeping various units were critical to the success of the Trust's work, in view of the huge costs that would be entailed in maintaining units of more than 40 breeds without their help. Provision of support for these breeders, including registration programmes, insemination services, a Show & Sale and the promotion thereof would, in and of itself, require a substantial organisation. It was his opinion, therefore, that this organisation should also be utilised for fund-raising and promotional purposes through a broadened membership structure, the revenue from which could be used to offset (or partially offset) those costs that were essential to servicing the core of livestock breeders and other essential parties. Rosenberg also held the view that a large membership had benefits in excess of the actual subscription income received. Many members paid under a Deed of Covenant, made additional donations, purchased raffle tickets and merchandise and provided a pool of help as stewards at the Show & Sale and on demonstration stands and, in total, provided circulation numbers to encourage further advertising for *The Ark*.

A disinterested observer might wonder if the 'technical' element is a little wary of the 'popular'. Both approaches have the same ultimate aim: to save endangered breeds. Perhaps there are those who feel the job should be confined to the experts, to an exclusive little club, rather than by appealing to the 'masses'. Perhaps there is just a hint of elitism, and a distaste for having to take into account the views and criticisms and apparently uncontrollable activities of a much wider group. Perhaps, too, the Trust might not have achieved its international leadership in its field if it had remained small and exclusive.

An organisation based on the idea of a large membership needs to have a structure which can service that membership, and Rosenberg was certainly capable of creating uch an organisation. Indeed he relished the challenge. The Trust had developed a mushroom' structure in the early stages — a large cap on top (in the form of a ubstantial administration) supported by a thin stick of membership. Over the years he stalk has thickened but the administrative cap is still too heavy for it: in 1992, Alderson calculated that each of the members was costing the Trust £3 a year more n membership support than their subscriptions covered. Growth in membership has een rapid. There were fewer than 100 members at the end of 1973; three months later here were 412 and by March 1977 there were about 2,500. In the 1990s there are now nore than 10,000.

SUPPORT GROUPS

Membership is not merely a matter of numbers: it is also one of composition and nfluence. In that respect it is important to look at a situation that arose spontaneously ind spasmodically in the mid 1980s: the development of support groups. Although they now represent perhaps only 8% of the total membership, they are an increasingly well organised and therefore influential minority in the affairs of the Trust and, again, that has inevitably led to a degree of discussion, if not actual conflict.

As the groups developed, Dr Richard Allan (co-opted to Council 1983—84) realised hat it might benefit the Trust's work if the groups were more organised. In 1985 he attended a crucial policy meeting held in October at Appleby Castle by invitation of Denis Vernon specifically to raise the subject. The first half dozen or so support groups had formed themselves a couple of years earlier and by 1985 they had increased to 22. Inevitably, by the informal nature of their development, there had been a certain amount of lack of communication with the Trust and no formal control over their activities. In 1986 Allan became chairman of a new Support Group subcommittee and, n that capacity, became also an ex officio member of Council. This was quite a significant move in that it gave the support groups a voice on Council. At that stage a structure of regional grouping was established and constitutions were formed, but the groups retained their independence and set their own standards.

Allan was succeeded as the subcommittee's chairman by Peter Titley and there is no longer an automatic place on the Council for the support groups. At the end of 1992, the chair was taken by Sue Dickinson.

Some of the groups did their fund-raising supremely well: for example, the Northampton group raised £14,000 in one year. The Dales group was another which made great efforts and put on ambitious events for the sake of rare breeds. Some on the other hand began to diverge from the Trust's policies, or at least to query them, and one or two simply went their own way. However, the Charities Act of 1992 laid down model constitutions for support groups, and this removed many of the contentious issues.

ADMINISTRATION

In the beginning, the administration of the Trust was carried out by RASE staff at the National Agricultural Centre, Stoneleigh, under the direction of Christopher Dadd. It was allowed the free use of the RASE's London office for meetings, for which no charges were made until 1974. Just before the Trust was officially launched as a charity in 1973, the administration was taken over by Peter Hunt, its newly appointed Secretary, on a part-time basis from his own home in Abbots Langley, while he also launched various fund-raising schemes. He continued in this multiple role until the pressure of other work became too great to combine with the rapidly growing Trust's demands and he withdrew in 1978: formally his retirement from the Trust took place on 1 December 1978, and he had generously given due warning of its imminence as early as June that year. His role in the formation and development of the Trust had been a crucial one.

As a temporary measure it was agreed that Alderson's company, Countrywide Livestock Ltd., should take over administration of the Trust and that the Hon. Treasurer, Vernon, should act as company secretary. For a brief while the Trust's papers were held at Michael and Marianna Rosenberg's Ash Farm in Devon. Here *The Ark* was edited and the Combined Flock Book maintained in the presence of an aging cockatoo and Rosenberg's model railway.

The Trust's headed notepaper of the period reflects some of the office's comings and goings. In a letter of 18 January 1979, the Trust's chairman Richard Cooper inked out Hunt's Abbots Langley address, altered Alderson's address from Colerne, Wiltshire, to Haltwhistle, Northumberland, and typed in his own address for replies, adding a plaintive note: 'I think I ought to have some respectable notepaper!'

Then in March 1979 R.L. (Bob) Smith, who was a business management consultant in Northumberland, was appointed as administrative officer to succeed Hunt. The Trust's office services continued to be supplied by Countrywide Livestock Ltd. in Haltwhistle, Northumberland, from a building standing alone plumb in the middle of Market Place.

The Trust was in effect administered from someone else's offices and making use of someone else's staff (one and a half secretaries, according to Alderson at the time), and being run by Smith as administrative officer and Alderson as its part-time, salaried Technical Consultant. Alderson was living in Bonnyrigg Hall, by Hadrians Wall, a wild place cut off from time to time by deep snow.

During 1979 the letterhead was sorted out: the address became simply Market Place, Haltwhistle, Northumberland. The town was just over the hill from Appleby Castle, the company headquarters of Ferguson International Holdings, whose chairman at that stage was the Trust's treasurer, Denis Vernon.

Smith's term of office was brief and ended abruptly in the summer of 1980. Vernon generously agreed to take the Trust's bookkeeping and membership records into his own offices at Appleby Castle as a temporary measure and a subcommittee (Rosenberg, Vernon, Dadd, Hindson and Wheatley-Hubbard) was formed to find a Secretary. It advertised the position and considered more than 400 applications. After 18 applicants had been interviewed, a short-list of three candidates was presented to the Council in September 1980.

There was a considerable difference of opinion among the Council members and no consensus could be achieved. One of the main problems was that all of the candidates had other interests, which they wanted to continue, and all wished to operate from their own present locations. The other problem was that it was difficult, if not impossible, to agree on a job description as there was no actual office in existence. The meeting adjourned in a certain degree of disarray. Forwood, who had been present, soon telephoned Rosenberg to say that he had been asked by various Council members to find out if Rosenberg would undertake to establish an office on behalf of the Trust and to get it up and running in the capacity of Honorary Director. It was felt that it would be easier to employ a Secretary if there was an established and functioning office. Rosenberg considered the matter carefully and then agreed; he began to look for suitable premises.

From the start, his search focused on the National Agricultural Centre at Stoneleigh, home of the RASE and venue for the Royal Show. He visited several organisations based at the NAC and found that surplus space and staff were available at the Shorthorn Society's permanent offices there. Rosenberg discussed the Trust's situation with the Shorthorn Society's Secretary, John Wood-Roberts, who had initially been approached by the auctioneer John Thornborrow. The result was that the Trust moved in during January 1981. The society agreed to create a small room for

Rosenberg's own use and it was only necessary for the Trust to hire a single employee, Mrs Eleanor Hewitson. All other work was carried out by the society's own employees on a part-time basis.

Administration of the Combined Flock Book, *The Ark* and the Semen Bank programme were relocated to Stoneleigh from Alderson's offices in Haltwhistle. Hilda Powell, of the Shorthorn Society, handled applications for semen purchase; registrations and preparation of the Combined Flock Book (and, later, the pig registration scheme) were undertaken by Cathie Church (née Calvert), who also worked on the Shorthorn Herdbook. With *The Ark* once again under his own direction, Rosenberg decided that it now deserved professional production and he engaged Alec Paris Publicity in Banbury for the purpose; he was also lucky enough to obtain the help of Mrs Pat Cassidy.

Meanwhile the membership records and the Trust's bookkeeping had been transferred from their temporary home at Appleby Castle. Since the days of Peter Hunt the membership records had been maintained on a series of 3 x 8 in. cards with stencilled or handwritten addresses at the top and it was immediately obvious that the system needed to be changed, urgently. Rosenberg engaged the services of Protext, a data processing firm based in Ely and owned by Jean and Ray Allen, who undertook to transfer the basic data to computer and to produce envelopes for the monthly mailing of *The Ark*, together with subscription renewal notices. The system would also be applied to the Deed of Covenant records, which had been sadly neglected since Hunt's departure.

Rosenberg soon found that the transference of so much data to disk was a task which stretched beyond normal working hours. Fortunately Mrs Rosalind Ragg volunteered her assistance and spent endless hours, often at weekends, helping him with the work. She was sometimes accompanied by her Great Dane, Penny, who was almost too large to fit into the little 8 ft. x 10 ft. office.

When the transfer of the membership records to computer was nearly complete, Rosenberg was anxious to see what the new system could do and he asked Protext to produce an enclosure, to be sent to all fully paid Life Members, requesting a contribution to the Trust's Livestock Support Fund. To the elation of all concerned, the response was such that donations far exceeded the cost of transferring all those records.

After several months in these new quarters, Rosenberg suggested that the appointment of John Wood-Roberts as Secretary of the Trust as well as of the Shorthorn Society would be an obvious and happy combination. The Council approved his suggestion and Wood-Roberts took up this additional responsibility on a formal basis. He was a man with a solid and practical farming background, with a great love for horses, and his considerable administrative ability and loyalty became invaluable to the Trust.

Space had already become a problem by 1982 and the Trust therefore leased additional facilities from the South Devon Cattle Society in an adjacent building so that it could hold committee meetings 'in house' and also provide room for Pat Cassidy's work in producing *The Ark*.

Wood-Roberts was generous with his services for several years but by 1987 he was beginning to experience something of a conflict of loyalties. Trust work was taking up more and more of his time but his first commitment remained to the Shorthorn Society; he would soon have to reach a decision about who really was his employer. In addition, the Trust was rapidly outgrowing the space available in the Shorthorn offices.

The Trust's first thought was to see if the Shorthorn building could be extended to accommodate that growth but this did not prove feasible, a judgement which was agreed in March 1987. After negotiations with the RASE's chief executive, John Hearth, the Trust rejected an offer of premises at East Lodge but decided instead to build its own new offices. There was a quotation of £80,000 on the table and, as had become the custom, there was a willing funder, Rosenberg, in the background. At last, the Trust had a home of its own. The new offices added to the presence of the Trust at the NAC, where a permanent Rare Breeds Pavilion had been opened in 1983.

*

The second half of the 1980s saw major changes in the executive staff. A new post was created in 1986 when Alastair Dymond was seconded from the RASE to become the Trust's chief executive. When Wood-Roberts finally withdrew in 1988, Robert Terry succeeded him as company secretary and administration director. Terry remains responsible for staff and for the office.

By 1991 the Trust's technical programme had increased so much that it was necessary to appoint a Field Officer as a technical assistant. The once part-time Technical Consultant had become the Trust's Executive Director in 1990 with a staff of 13. Details of the staff in the Trust's three main departments (administration, technical and publicity) are given in the Appendix.

PHASES OF GROWTH

Dominant personalities have shaped the growth of the Trust and some names in particular stand out. The influence of Zuckerman and Forwood in the early days, and latterly Barber, has been acknowledged. But three other figures stand out; Christopher Dadd, Michael Rosenberg and Geoffrey Cloke. Christopher Dadd was both the engine-room and the steering gear during the 1960s and early 1970s; Michael Rosenburg led by example, effort and experience from the early 1970s to the mid 1980s; and Geoffrey Cloke's enthusiasm and initiatives carried the momentum into the early 1990s.

The Trust's development can be divided quite easily into distinct phases, during which there have been some outstandingly fruitful periods. The first, preparatory phase began with the joint meetings between ZSL and RASE back in the late 1960s until the acceptance of the Trust Deed and acceptance by the Charity Commission in 1973. This was the period in which Christopher Dadd, Idwal Rowlands, Ann Wheatley-Hubbard, Peter Jewell, John Bowman, Sir Dudley Forwood, Joe Henson and Bill Stanley played critical roles. Without them, the Trust would never have been formed.

The second phase was one of rapid development and growth, an expansion triggered to a great extent by the two paid part-time consultants (Hunt and Alderson) and the paymaster and major driving force, Michael Rosenberg. During this phase, which lasted right through the 1970s, the TAG initiated many technical projects and the SSC was much more of a working body than Council. The vintage year of 1975/6 saw the birth of *The Ark* and the Combined Flock Book, the Show Demonstration programme, a series of Breeders' Workshops, the first Show & Sale and the first joint Scottish symposium at Edinburgh Zoo, the establishment of criteria for approved centres, the importation of boars from Australia, and the establishment of rare breeds acceptance procedures. Another brace of years in this period, 1979/80, saw the Trust's first major policy statement and a policy seminar at the NAC which generated many new ideas; there was also an FAO meeting in Rome at which it was clear the Trust's work was well ahead of any other country's in the genetic conservation of domestic livestock.

The third phase was one of consolidation during the first half of the 1980s: the innovations had been made and now it was time to form a solid basis and make sure that everything worked properly. It was also a period of strong chairmanship from Rosenberg, Wheatley-Hubbard and Cloke. The Trust's finances were put on a sounder basis and various systems were put in place to ensure the smooth running of the Trust's work.

During the latter half of the 1980s Rosenberg, for personal reasons, withdrew from his daily involvement in the Trust and perhaps people lost their way a little without his sound organising abilities and effective leadership. Ironically, the more spontaneous than analytical approach that Alderson had personally preferred seemed to take over: development was more random and perhaps lacked cohesive overall strategies. The membership, of course, had become much larger and was growing all the time. More and more people came in with their own ideas and, inevitably, conflicts developed between the long-standing and the new members. The Support Groups were growing, for example, without being clear about their own function in relationship to the Trust.

Executive Director

Alderson had been gradually pulling out during the late 1980s, with the intention of retiring from the Trust, and this served to compound the sense of a vacuum created by Rosenberg's withdrawal. Alderson had transferred the technical department to his own offices at Droitwich and, after so many years of close involvement, he now had a little time to stand back and see the Trust from a more distant perspective.

In 1990, he was approached to work for the Trust on a full-time basis for the first time, now as its Executive Director. Like others who had served the Trust long and well, he was loth to leave it in other hands which, for all he knew, might radically alter what he had helped to create. He therefore accepted the post, and helped to stimulate a renewed phase of growth while at the same time imposing standards and procedures which he felt had slipped in the meantime. 'I sympathise with the British government,' he said, 'who in 1992 were simultaneously trying to encourage growth in the economy while at the same time holding down inflation. They are two contradictory things and I felt pretty well the same about the Trust.'

And, like an unpopular chancellor of the exchequer, he found that many resented his 'disciplinary' measures. He expanded the technical programme, and the annual Show & Sale; re-established a full centrally organised programme of breeders' workshops (which had been devolved to the support groups), and became involved on both the political and the international fronts. To help him in these new goals, Peter King was appointed as a Field Officer in 1991.

At this stage the balance of power between Council and Executive swung again. Alderson produced an article from a charities magazine and quoted the following paragraph:

> The day-to-day business of all charities is conducted by the professionals. Policy decisions are made by the trustees and the chair of the trustees is theoretically the professionals' employer. There needs therefore to be a good working relationship between the trustees or council and the professionals. Most intelligent trustees have faith in their appointed directors and leave them to get on with the job in hand. Make no mistake: directors run their charities and indeed prepare the agenda for trustees meetings. In other words, they decide in the main what subjects are for discussion and what are not. That is why they get paid!

1990/91 might prove to be another vintage period: it heralded a new growth phase and also major international involvement (with the creation of Rare Breeds International) and recognition at home that the Trust had come of age. In 1991, the first major policy statement since 1979 was drafted and approved by Council.

FUND-RAISING

Today, the main sources of funds can be divided into three categories: membership subscriptions, interest and dividends from investments, and donations or sponsorship (specific or routine). Each represents roughly a third of the Trust's income. Income increased constantly, although at a varying pace, until the effects of the severe economic depression were felt in 1991 and 1992.

The Trust would never have got off the ground in the first place without the fund-raising efforts of Peter Hunt, its first Secretary, and success was assured by the efforts of Rosenberg. It owed a huge debt to the grant-giving trusts that Hunt successfully approached, many of which are still involved today. Perhaps the most significant (in terms of timing, generosity or faithfulness) have been the Ernest Cook Trust, whose pump-priming paid the Technical Consultant's fees for three years in the early days; the MacRoberts, to which the field study centre at Lower Midgarth owed its refurbishment; the Esme Fairbairn, which gave generously towards the pig importation programme in the 1990s; the Balerno, for its long support of the AI and Semen Bank scheme; and the Sun Alliance for its sponsorship of the Show Demonstration programme. The Trust has also been ably served by its honorary treasurers — Sir Dudley Forwood (1973—5), Sir Richard Cooper (1975—7) and Denis Vernon since 1978.

For a while in the 1980s there was a fund-raising committee chaired by Cooper; its members included Christopher Dadd and Michael Rosenberg, with all their contacts, Vernon as the Trust's treasurer, and Christopher Marler. In late 1982 Marshall Watson was appointed as Fund Raiser and he remained in that post until July 1984. By 1985 Cooper wanted to review the position of his committee: it had been agreed to appoint, from 1 January 1986, Alastair Dymond as the Trust's Chief Executive, with fund-raising as one of his major roles.

Chief Executive

The possibility of Dymond's secondment from RASE had been discussed between Cloke, as Executive Chairman, and John Hearth, chief executive of the RASE, in 1984. Dymond was Deputy Agricultural Director at the NAC and, before taking up his new appointment with the Trust, he was invited to attend a policy meeting at Appleby Castle in October, 1985, to make a presentation about his plans for the Trust. Ambitiously, he said that he envisaged a future membership of 20,000 or even 30,000. As well as increasing the membership, a major part of his remit was to raise funds (he later said that the success of his employment would be measured by whether his fund-raising significantly exceeded the cost of employing him), and much of this would be done through the Show Demonstration programme. This would all be in tandem with his administration responsibilities.

The Appleby meeting listened intently. They were well aware that Rosenberg, with a combination of expertise and funding, had given the Trust a high reputation for quality in its activities, especially through the Show Demonstration programme and the hosting of fund-raising social events. They wanted to maintain that feeling for quality but on a less extravagant (and less expensive) scale. That was one of the challenges faced by Dymond.

Alastair Dymond had become a member of the Trust in 1973, the year in which he had become the NAC's farm manager, and his interest in unusual breeds stretched back to his time as assistant farm manager to Sir Walter Burrell, who ran an ancestral flock of Jacob sheep. Dymond's most serious involvement with the Trust in the early days arose through the Show & Sale — the RASE entered sheep and cattle over the years. He remembers the early events well: the first Show & Sale was 'tiny, one shed and a few pens tied together with string'; the second, for which he was a sheep judge, was not much bigger. When Rosenberg took charge of the third event in 1978, he invited Dymond to become sheep steward. He took responsibility for organising the sheep until 1985.

In 1984 Rosenberg's father died and he had much less time to devote to the Trust. Cloke came in during 1985 to fill the gap but Rosenberg had already suggested that a chief executive was needed. Cloke began negotiations with Dymond and they finally reached an agreement whereby Dymond was seconded from the RASE: his salary continued to be paid by the society from 1986 to 1990. He began to be involved from the autumn of 1985, attending shows with Rosenberg in connection with the Show Demonstration programme.

His objectives for the Trust were quite clear. He felt that the membership must be expanded considerably — initially to some 10,000 (at that stage it was about 6,000). He wanted to instigate a proper accounting system, making sure that quotes were sought before ordering and so on. In his first show season he cut the budget to £20,000 and achieved sponsorship from Sun Alliance, which reduced it to £10,000. He calculated that it would be cheaper and more efficient to instal the Trust's own computer system, rather than using outside agencies for wordprocessing and membership organisation, and also for accounts which at that stage were still done by hand; indeed, it could also computerise the Combined Flock Book and the trading company. The system cost some £25,000 to install and the cost was written off after five years.

In 1990 his term of secondment from the RASE came to an end. Changes to the staff structure of the Trust were in the wind and Alderson came in as Executive Director. Only one year before the position of Field Director disappeared in further changes. Dymond left the post in March 1992 to set up his own advisory service for small livestock undertakings.

MANX LOGHTAN

CHAPTER FOUR
EARLY VENTURES

SUFFOLK MARE

BRITISH LOP SOW

CHAPTER 4:
EARLY VENTURES

The freshly formed Trust of 1973 was bursting with ideas as to how it could meet its self-appointed task. Inevitably mistakes were made: they were pioneers, after all. Not one breed has been lost since the Trust swung into action, and many have found a real and lasting new lease of life. That is the only real criterion by which to judge the Trust's success.

THE LINGA HOLM STORY

One of the old Working Party's members was Captain Pitman, representing the Fauna Preservation Society, and it was he who brought the new Trust's attention to the FPS's Orkney Sheep project. They were particularly worried about North Ronaldsay, the only island where the purebred native sheep still survived on its mainly seaweed diet. The FPS felt that the Trust, involved with domestic animals, was now the appropriate organisation to conserve the sheep of North Ronaldsay. The Trust agreed in principle that it was important that this remarkable breed's future should be insured by forming a separate breeding population in case of some disaster befalling the isolated flock on North Ronaldsay.

The FPA supplied a report, by their member Baxter Cooper, which suggested three islands that might be available as sanctuaries for the sheep, including Kili Holm at about £1,000 and Linga Holm at £2,600. Joe Henson was asked to take a trip to the Orkneys and Deryk Frazer carefully organised a coincidental official Nature Conservancy trip (in relation to seals) so that the two men could look at various islands together. During the week of 15 October 1973, they went to North Ronaldsay to look at the sheep and talk to local experts such as Mr and Mrs Muir, who lived on the island, Miss Bullard of the Orkney Field Club, who provided some botanical background, and Bill Carstairs, veterinary officer for the northern isles, who lived on Sanday. The Muirs and Carstairs would remain as important Trust contacts for many years.

Frazer and Henson reached an agreement with Carstairs that he should, on the Trust's behalf, purchase about 70 ewes, 70 lambs and 10 rams, of various colours, at £7 a head, from a cull on North Ronaldsay during December. The next step was to find somewhere to put them and they began a survey of the islands.

Unfortunately one of the three islands recommended by Baxter Cooper had already been sold but there were other possibilities. Henson remembers that first trip vividly: they flew in a small aircraft piloted by Andy Stewart, with six other fee-paying passengers on board, heading for Stronsay. Stewart decided they could have a good view of some of the islands from the air and, slightly to the consternation of the rest of the passengers, he pitched and swooped over likely places. They landed safely at Stronsay and planned to make an equally pitch-and-swoop journey to the islands by boat but Frazer, with his naval experience, said the sea was far too rough for such an escapade! It turned out that several of the possible islands were tiny holms where, at high tide, there would be barely enough dry land for the sheep to huddle together.

The two men met Sam Cooper on Stronsay, at his home, Midgarth; he owned the island of Linga Holm just across the sound and he took them to view it. They both agreed that this was the best choice: it was 142 acres, rising to its highest point at the centre so that there was shelter from winds in any direction; there was ample drinking water. There were only a couple of derelict, roofless buildings and there was not a huge amount of kelp but otherwise it looked satisfactory. Then they spotted a tumbledown ruin on the Stronsay shore opposite the island and Frazer gave Henson a nudge: 'Get him to throw that in with it,' he said.

Baxter Cooper's report had suggested that Linga Holm could be bought for about £2,500 but Sam Cooper denied all knowledge of such a price and insisted that his asking price was £40 an acre — which would mean more than double that figure. But, yes, he could also sell them Lower Midgarth, the ruined house on Stronsay, with a bit of land around it.

Frazer and Henson returned home with a predicament. They had instructed Carstairs to buy some 150 sheep on behalf of the Trust but they had nowhere to put them. Christopher Dadd talked privately with Lord Dulverton at Batsford Park, Moreton-in-the-Marsh, and his cousin, Captain Andrew Wills, who owned the three Crowlin islands. Wills told him, however, that none of them were likely to be cleared of crofters' sheep for perhaps two years, but that he was very interested in the North Ronaldsay sheep situation and would look into it in due course to see if the interests of the Trust might coincide with his own future plans for one or more of the Crowlin islands.

The problem remained that the Trust had authorised a sum of only £2,500 for the purchase of Linga Holm and the asking price was in the region of £5,500. Trust members were divided: the matter was urgent but should they buy it now, before the price rose? Or was the price already an inflated one, and should they really become involved in property, as such, anyway? A down-to-earth businessman on the committee pointed out that it would take all the Trust's slender funds and that no organisation in its right mind would risk bankrupting itself on one project. Henson said cheerfully that they would be able to recoup such an outlay within the year. Dadd and Alderson (newly appointed as Technical Consultant) were among those strongly in favour of buying the island and Dadd held out the hope that the World Wildlife Fund or some other charitable organisation might be able to bridge the difference between the Trust's original budget and the asking price. But time was pressing and in they end they decided to go ahead: they would buy Linga Holm and Lower Midgarth and worry about their funds later. The negotiations for Linga Holm were completed in May, 1974, and it turned out to be one of the best public relations moves they could have made.

Suddenly Peter Hunt's fund-raising had an urgent target. He had included the North Ronaldsay sheep in his initial appeal document to the grant-giving trusts but had allowed only £2,000 for the purchase of a sanctuary for them. He now arranged a very much 'live' press conference in London's Fleet Street on 29th July, 1974. Henson, with his typical flair for publicity, borrowed some tame North Ronaldsay sheep from Riber Castle (those on the island itself would have been completely unmanageable) and paraded them on leads down Fleet Street for the ideal emotive photo-opportunity: odd-looking primitive sheep that lived on seaweed on an Orcadian island, now set against the background of St Paul's. It certainly worked: the Trust achieved some 500 column inches of media attention, and Hunt managed to recoup the outlay with interest — they raised considerably more than they needed to buy the island and to repair Lower Midgarth. That tumbledown building, once planning approval had been received in March 1975, would be transformed into a field centre and renamed MacRoberts House after the grant-giving trust which had given it so much support. The venture had highlighted the benefits of fund-raising for a specific project rather than making general appeals.

The small flock of 178 sheep (seven rams, 114 ewes and 57 lambs) was separated from the main stock on North Ronaldsay and successfully transported to Linga Holm as the reserve breeding group. At the same time another 108 sheep were dispersed to flocks on the Scottish mainland and in southern England and the Channel Islands as added insurance, where they continued to breed, replenished from time to time by more sheep from North Ronaldsay and Linga Holm. Meanwhile the Linga Holm flock settled in well. In October, 1974, Deryk Frazer accompanied the Trust's president,

Lord Cranbrook, to Linga Holm and they reported that the sheep were well and were grazing both grass and kelp. A veterinary report in June, 1975, commented on an excellent lambing, a very good group of rams, plenty of kelp for them, and suggested there was ample scope to enlarge the flock. Within two years there were more than eight hundred sheep on the islands and a cull was overdue.

Any Trust member who has visited Linga Holm has good memories of the adventure — and adventure it invariably has to be, in the face of atrocious weather and stormy seas. The working party learned from painful experience that visits after August were risky. On Alderson's first visit in October, the seas were so mountainous that the lobster boat, which had been hired after much persuasion, spent most of the journey standing on its bow or on its stern. Even when the visits were fixed in the summer, there were times when the party became stranded on the Holm by winds and strong currents, or tried in vain to take shelter from drenching rain. However, spirits always remained high.

Alderson's report to the Advisory Committee in 1975 described his visit to Linga Holm on 22nd October and gives the flavour of the place:

1. Conditions: Visibility good, cloudy but no rain; wind fresh; sea rough.
2. Transport: The sea was too rough for the small boat, and a lobster boat was hired.
3. Buildings/Facilities: The sheep pens are small and inadequate to deal with the flock of North Ronaldsays on the island, both in terms of space and the height of the walls. There is a house (2 rooms plus kiln) and a store, both roofless and a small quay of stone slabs.
4. Sheep: (a) There were appreciably more than 300 sheep on the island. An accurate count was not possible but the approximate breakdown was 105 adults, 70 shearlings, and 160 lambs.
   (b) They were generally in very good condition and active. There was quite a bit of loose dung and one ewe appeared to be scouring. The lambs had not been weaned and several ewes were still suckling a good pair of twins.
   (c) The sheep were not shorn this year but out of the total flock only three ewes had started to shed their wool.
   (d) The group of rams had split up and rams were mixing individually with the flock in readiness for mating.
5. Forage/Seaweed: There was an abundance of grass and the sheep were making very little impression on it, although a good proportion of the flock was grazing in the interior of the island. I was unable to inspect the shore at low tide so that the kelp was covered, although the wrack was exposed.
6. Seals: In excess of 30 seal pups had been born at the time of my visit.
7. Recommendations:
   (a) The sheep (excluding lambs and hoggs) should be sheared in 1976.
   (b) The flock should be reduced in size in the autumn of 1976, to about 330 adult and shearling sheep.
   (c) The flock should consist of approximately 16 rams, 130 ewes, 185 hoggs/shearlings, and the current crop of lambs.
   (d) The annual cull should consist of about 175-185 sheep. No lambs should be culled.
   (e) A range of colours should be maintained by using rams of the following colours: white and grey 13, black and brown 3.
   (f) Apart from maintaining the correct colour ratio selective culling should be carried out only to remove deformed animals.
   (g) The handling facilities should be improved. Approximately 450 yards of fencing will be required.

(h) Mr Carstairs should be asked to analyse dung samples for nematodes and liver fluke, and the Orkney Field Club should be asked to report any incidence of fluke snails.

8. Financial implications: The output of wool and sheep for slaughter should exceed £1,600, thus giving a total income (including the seals) in excess of £2,000. Against this must be set the cost of shearing, haulage, Mr Cooper's retaining fee, and my own expenses, which should not exceed £600, leaving a margin of £1,400 plus. Any income from the house would be extra.

In 1977 Peter Jewell proposed the idea of a student undertaking a study of the sheep on Linga Holm and North Ronaldsay, and the case studentship reveals a few more snippets about the island. By 1977 there were 828 sheep (203 rams, 371 ewes and 254 lambs) on Linga Holm and a cull was initiated. The feral flock was available for research, and the shore of the island offered a unique opportunity to study seaweed as a food resource for the sheep: the shore was extensive and the seaweed was exposed at low tide in abundant stretches. The sheep also used the island's pastures. The Trust's house on Stronsay, opposite Linga Holm, offered free accommodation for the student, who could also have the services of a warden/boatman (Sam Cooper) engaged by the Trust to assist with the management of the sheep. A field hut was being built and a Mk.II Zodiac 12ft dinghy with a 10HP outboard motor was being purchased for use by the working party.

Alderson visited Linga Holm regularly, and Jewell also made regular trips, holding a personal grant from the Leverhulme Trust for the expenses of his own travel to Orkney and St Kilda to study the sheep in the period 1978-80. Alderson remembers with joy the flight to Stronsay on an old DanAir light plane, going down low over Linga Holm on the way and banking over it so that it was almost possible to complete the inspection from the air, then landing on a cow-pasture airstrip on a headland opposite Midgarth in strong gales and lashing rain. He continued to help for a few years and remembers banqueting on wild mushrooms and rabbit with Don Clinch, and even on one occasion making the most of the accidental death of one of the sheep for a mouth-watering meal once it had been skinned by Jewell.

Alderson's confidential report in September 1982 gave an interesting brief history of the island of North Ronaldsay itself and the intriguing system of management there by a Sheep Court under regulations drawn up in 1902. In 1910 there had been 442 human inhabitants on North Ronaldsay but the population had declined steadily until by 1982 there were only fourteen able-bodied men available to work with the sheep and repair the wall system (twelve miles long and six feet high). The wall confined the sheep to the shore for nine months of the year, where they subsisted on seaweed, but allowed three months on grass with their lambs. The available labour was by then scarcely adequate to maintain the wall or to gather the sheep at traditional 'pundings' for dipping, ruing etc. The condition of the sheep on North Ronaldsay had deteriorated by the 1980s and with a population of more than 4,000 animals, and increasing, it seemed likely that the island (three miles long and a mile wide) was overstocked. There was a real threat to the future of the flock. In contrast, the Trust's flock on Linga Holm (at that stage about 175 ewes and 25 rams, plus followers) was in good condition: it had been managed annually by a working party and supervised for the rest of the year by the veterinarian Bill Carstairs, then living on Sanday, with the help of Ian Cooper, son of Sam Cooper (the original owner of Linga Holm), who kept an eye on the flock all year round.

However, the problems on North Ronaldsay served as a warning: the Linga Holm grazing pattern might also cause difficulties. Sam Cooper had remarked, in the first year, that 'one would think there was a dyke around the island — the sheep never venture inland more than a few yards' and they were clearly happy with their seaweed.

The following year, though, they began to favour the grass and by 1982 the herbage was heavily grazed, which possibly indicated that the breed's physiology was altering. Steps needed to be taken. For example, the Trust is now looking for markets for the special wool of the North Ronaldsay from its native island to encourage the keeping of that flock, and is investigating the idea of creating a market for gourmet meat from the breed. It also reacted quickly when severe storms in 1993 brought down more than two miles of the wall that separates the shore from the interior of North Ronaldsay: it gave assistance and financial support to employ experienced dry-stone wallers to repair the damage. Another recent idea was to purchase another island and the Trust put in a sealed bid of £30,000 for one but it was not accepted, though it was the only bid received.

The Linga Holm working party has done wonders for the sheep. It was first set up in 1977 and right from the start its natural leader was Ken Briggs, aided and abetted by his wife Nancy, whose cooking restored vitality to many a working party member. Briggs was an ideal choice as leader of the working party: he is a practical farmer with a sense of adventure. During the years since his first visit in 1977, he has developed the fencing and penning system, negotiated with neighbours for essential and often urgently needed equipment, and visited North Ronaldsay to establish a working relationship with the Laird. With growing experience he has learned the vagaries of the Orkney winds, tides and currents, and developed the ability to know which beach on the Holm offers the best opportunity for landing in various weather conditions.

Others have become involved over the years — men like Don Clinch (part of the Show Demonstration programme) and Howard Payton, and more recently Michael Dickinson and Eric Freeman (a breeder of Gloucester cattle). There is a high demand to become a member of the working party on Linga Holm.

BOAR IMPORTATION PROJECT

At its meeting on 5 August 1975 the Council had considered the possible importation of boars from Australia. The subject arose during discussions about the possibility of using AI and establishing a semen bank for pigs, in the same way as the MMB had co-operated with the Trust for cattle. It was recognised that most of the country's pig breeds were critically low in numbers, since the pig industry had switched entirely to intensively reared Large White and Landrace combinations and it had become rare indeed to see an outdoor herd of the old, hardy, coloured breeds. The tumble had been a relatively rapid one and the main problem now was the lack of bloodlines in the national stock of breeds like the Tamworth and the Berkshire. Fortunately those two (among others) had been widely exported in their heyday to stock the old colonies and there was therefore a useful source of genetic material in Australia and New Zealand, where they had been popular among dairy farmers to make use of dairy by-products.

The Trust decided to import some Berkshire and Tamworth boars from Australia and Jack Howlett of the British Livestock Company flew out to select the best of the stock. On 24 August 1976 three Berkshire and three Tamworth boars landed at Heathrow. They were taken into quarantine at the MAFF centre in Plaistow.

The project was financed by the Trust (it had set aside a sum of £3,000 towards the project in January 1976) in association with the NPBA, and a detailed breeding programme was drawn up in conjunction with the relevant breed councils. Time was at a premium: the boars' useful breeding life would be perhaps 12-18 months, and it was essential that they should service as many sows from as many different herds as possible in that time. After their quarantine, they were placed on selected farms in different parts of the country.

There were bound to be some problems, and some internal wrangling over the administration of the agreed breeding programme. Several Council members felt that the NPBA was the proper body to supervise the programme, in co-operation with both the breed societies, rather than the Trust's Technical Consultant though the boars officially belonged to the Trust.

The Tamworth boars (Jasper, Glen and Royal Standard) flourished but unfortunately two of the three Berkshires died, leaving Ambassador to create a powerful impact on his own. Lessons were learned and the Trust was well prepared for its next importation in the 1990s, when a project to import Berkshire, Large Black, Saddleback and Tamworth was developed.

SEMEN BANK AND AI PROJECT

Even before the Trust was officially formed, its Working Party had persuaded the Milk Marketing Board to incorporate some of the rare cattle breeds into its 'museum' semen bank and AI scheme. In the spring of 1975 Alderson proposed that a Trust semen bank should be created for each of the Gloucester, British White, White Park, Shetland, Longhorn, Northern Dairy Shorthorn, Red Poll, Kerry and Dexter cattle breeds, in conjunction with the MMB and (it was initially hoped) ABRO, though the latter did not become involved in the end. The MMB, through the efforts of Keith Cook, co-operated fully and helped the Trust to obtain a grant from the Balerno Trust at £200 per bull for the bulls which had contributed 100 straws of semen to the MMB's Bank of Genetic Variability (the museum bank). In June 1975 the Trust launched its new cattle AI and semen bank project at a press conference. (Lord Balerno, incidentally, chaired the first joint Scottish Symposium on conservation, in Edinburgh.)

The project went well under the guidance of Alderson and by January, 1977, the costs to the Trust were working out at about £420 per bull, against which there was the Balerno grant of £200 and (over a four-year period) estimated semen sales of 75 straws from each bull totalling £178 — giving a small deficit of some £30 a bull, with another 100 straws from each in storage.

It is interesting to look at that original list of included breeds as it reflects one or two personal concerns at a time when decisions on which breeds should be included in the Trust's categories were not well defined. The Northern Dairy Shorthorn, for example, was included after persuasion from Alderson. The Irish Moiled was not included: there were too many question marks about it at that stage and the Trust was not sure of its purity. The Dexter cattle society, it was found, already had its own programme and there was therefore little point in including it in the Trust's semen scheme. The Gloucester had strong support from Robin Otter, Joe Henson and Peter Jewell, and the British White from Bill Longrigg and John Cator. In 1975 and 1976 the inclusion of different breeds or groups of livestock was determined by ad hoc discussion within the Council, though some standard formulae had been suggested in mid-1975 based on four criteria: vulnerability, purity, historical importance and recognition as a separate breed (see Acceptance Procedures, later in this chapter).

The Balerno grant ended officially in September 1978 but Keith Cook's diligence resulted in continuing support for the rare breeds semen bank, at first from the Balerno Trust and subsequently from other sources. Cook had given great service to the Trust throughout and continued to do so quietly until his retirement from the MMB in 1991/2; he was co-opted to Council in 1993.

By 1978, talk about collecting semen from a Longhorn bull came to nothing as the breed's society, like that of the Dexters, was involved in its own major project and there was little point in duplicating its efforts. However, the Trust is always prepared to collect semen from neglected lines of Dexter, Longhorn, British White and others.

By October 1980 the project had included eleven bulls from six breeds; 845 straws of semen had been used and the current price was £5 per straw (plus £2 consignment). There were criticisms that the potential cost of storage of the semen was too high and that insufficient bulls had been made available, but Alderson was able to point out that the current storage cost was only £10 a month.

He also talked about the question of a bull-rearing incentive with the idea of contract mating and then providing a grant for the rearing and (if it was a suitable bull) the subsequent breeding and collection of semen from it. And he talked about the possibility of embryo storage and transfer, an issue which has been considered from time to time since then, but the Trust has so far concluded that the cost benefit of embryo storage and transfer is not sufficiently attractive in comparison with semen collection and AI, nor is the technique so well established.

The current advice given by the Trust regarding embryos is that the Trust's main objective is to increase the population size of rare breeds and that, as the effect of any single family increase is diluted in the next generation by mating with unrelated bulls, it should put no restriction on the number of embryos from any single animal. There are dangers, but the main priority remains the maximum increase in population size. In principle, the cryopreservation of genetic material as semen, and also as embryos if necessary, is the ultimate insurance against the disappearance of a live population of a breed and against the natural inclination of breeders to select within their breed and to change it.

By March, 1983, the Trust had collected semen from 25 bulls representing six rare breeds. Apart from the straws stored in the semen bank, 1,400 had been used for artificial insemination, which meant that the best of the breeds were able to pass on their genes in both purebreeding and crossbreeding programmes. The objective is to collect semen from 25 bulls per breed every four or five generations.

In 1986 there was a proposal from the Breed Liaison Committee that the Trust should start to collect semen, fairly urgently, from breeds which were not on its list but were peripheral, especially if they were being subjected to 'improvement' or integration with other breeds. This idea did take root. For example, the Devon is suffering severe introgression from the Salers, and the Lincoln Red from the Maine-Anjou, in both cases within official programmes. Some old foundation populations of traditional breeds are also at risk from importations of doubtful status. Geoffrey Cloke and Alderson were alerted to such a situation in Hereford cattle by Sue Vaughan of Wye Valley Farm Park. The old type of Aberdeen-Angus is probably in similar danger.

The MMB continued to provide a grant for the collection of semen from up to six rare-breed bulls per annum, at the old Balerno rate of £200 per bull, into the next decade. By the end of 1986 the Trust had 5,392 straws of semen available for AI, over and above those stored in its semen bank, and they represented more than 100 bulls. By 1993 the number of bulls had doubled and 25,000 straws of semen were either in the bank or available for immediate use.

The philosophy of the semen bank was a very important one in the conservation of rare breeds. In due course it was extended to pigs, and in principle to sheep and goats though so far this has only been carried out in practice with the Norfolk Horn in the very early days (when it failed). By the end of 1986, however, the Trust had already stored 71 straws of boar semen in its bank, while semen collected from two Bagot goats at Reading had disappeared.

The development of the semen bank for pigs was particularly important as an insurance for vulnerable breeds. Long-term storage was carried out by Masterbreeders through the good offices of Rex Walters and Paul Hooper, while an artificial insemination service for rare pig breeds was provided by Pig Genetics. The latter arose from a chance meeting between Alderson and Christianne Glossop at an annual general meeting of the Pig Veterinary Society.

The main reason for storing deep-frozen semen from rare breeds is as insurance against some disaster befalling a live population but there is another side to it. Breeders will always breed selectively and seek to change their breed to suit their own purposes, and it is important to retain a bank of genes that in effect encapsulates a breed at a given period (at intervals of four or five generations) so that, should it be changed in the future, the original breed's qualities can still be resuscitated.

Yet all this was put in jeopardy in 1988 when the European Community issued a Directive about Trade in Bovine Semen which, at one stage, threatened to disqualify from use any semen collected before 1990. That was an excellent goad to get the Trust involved politically in Brussels and Whitehall to fight against such bureaucratic interference — a role which it is increasingly taking up now that it has come of age.

SURVEYS

The first survey of rare breeds undertaken specifically for the Trust (or, at that stage, for the joint Rare Breeds Working Party) was that carried out by Bowman's two students, Charles Aindow and Mary Underwood, in the summer of 1970, their expenses (£50) being paid by RASE and Mrs Wheatley-Hubbard.

The ZSL had consistently stressed the need for a proper survey during the 1960s; Ryder-Davies had carried out one privately during that decade, and Alan Marsden had offered, in a letter in January 1970 to the MAFF, to make a survey of 'scarce' breeds. It is unfortunate that no organisation seemed to be prepared to back such a survey at a time when some of the breeds which were then critically endangered might have been saved.

Although the Reading survey was necessarily incomplete and sometimes factually incorrect, it did highlight some of the breeds that needed urgent attention and gave the movement a focus. When Alderson was appointed Technical Consultant in late 1973, he immediately published an open letter to the Trust's members saying that the first priority was to discover the existing situation regarding rare breeds. In 1974 he carried out a more thorough survey which would be the basis, in 1975, of the Trust's formal priority lists (published in *The Ark* in September 75). The new survey noted, for example, the decline in population of Wensleydale and Shropshire sheep, Middle White and Berkshire pigs, and Irish Moiled cattle.

Surveys have been repeated at intervals of three or four years since 1974 but more recently it has been appreciated that populations of endangered pig breeds can fluctuate much more rapidly than those of other species. Thus a panel of pig breeds co-ordinators was formed in the late 1980s, its main function being to act as an early-warning system for the breeds and to carry out an annual survey. Members of the panel were Jo Clarke (Berkshire), Viki Mills (Large Black), Frank Miller and later Jenny Brett (British Lop), Alan Rose (British Saddleback), Andrew Robinson (Gloucester Old Spots), Alan Black and later Caroline Wheatley-Hubbard (Tamworth), and the late Joan Staig (Middle White). Pru Rose did a valiant job in co-ordinating information. Eventually the panel was amalgamated with the Pigs subcommittee of the Breed Liaison committee under Black's chairmanship.

The report forms that are used for these surveys have been progressively refined. In recent years, it has been considered prudent that they should be as compatible as possible with survey forms used in, for example, the Global Animal Genetic Data Bank at Hannover and in EC and FAO surveys.

RARE BREEDS ACCEPTANCE PROCEDURES

Breed surveys serve a dual purpose. They provide early warning of breeds that are declining, so that they can be given greater support if they are already on the Trust's lists or can be brought on to a 'Watching' list if they are not yet seriously endangered. Secondly, information from surveys is used to determine the status of each breed within the Priority List that guides allocation of the Trust's support.

It was not until the 1974 survey had been analysed that the Trust had adequate information to make fine judgements about which breeds it should include in its remit and on what basis. They had been using a provisional list based partly on the Aindow survey and partly on ad hoc Council decisions depending on various members' suggestions or special interests, but by late 1975 Alderson was in a position to propose clear-cut procedures by which the Trust could determine whether it should recognise and support a breed and, if so, in what category of priority the breed should be placed. The priority list sought to evaluate the degree to which a breed was in trouble and took into account factors other than mere numbers.

In a summarised version of the proposals it was stated, first of all, that 'unless a breed can demonstrate that it is in a vulnerable state or that it possesses a distinct characteristic which cannot be found elsewhere, and that it is largely free from crossing with other breeds, it will not qualify as a rare breed deserving the support of the Trust.' The essential criteria were vulnerability and purity. Vulnerability implied possible extinction or an ineffective breeding population, and the factors which might create a state of vulnerability were: (a) low numbers (the Trust had been considering critical minimum populations of breeding females as 1,500 for sheep, 750 for cattle, 100 for pigs); (b) number of bloodlines (at least three male lines required, and inbreeding could be kept within safe limits with six lines); (c) decreasing population (breed amalgamation was also relevant here); (d) limited geographical distribution. Secondly, had the breed been important historically? Was it an 'ancient' breed (more than 500 years old)? Had it been an important 'foundation' breed or basal type that had been used to improve old breeds or create new ones? Was it closely associated with an important agricultural area? Thirdly, was it a separate breed or genetic type? Did it breed true to a recognised breed or type? had it been known before 1900? and did it have an accepted herd or flock book? The final document stating the Rare Breeds Acceptance Procedure was agreed in September 1976.

Acceptance was of fundamental importance for a breed: it was only on acceptance that it would gain the recognition and support of the Trust, and it was the support of accepted rare breeds that was the raison d'être of the Trust (a fact which, perhaps, is sometimes almost overlooked). But setting the procedure for accepting breeds as rare was only a start. A refinement of that basic acceptance was the allocation of breeds within a priority list which sought to evaluate the degree to which a breed was in trouble and took into account factors other than mere numbers. It took quite a while to agree on priority categories for accepted breeds.

Controversial breeds

The new procedure was bound to lead to controversy when strictly applied to the breeds and a few would fall foul of the criteria, sometimes after considerable debate, or even conflict between individuals with opposing views about a breed. That was certainly the case with, for example, Blue Albion cattle, which are still excluded today. Contention in this case centred around a probable break in the continuity of the breed, which seemed to have become extinct after decimation during the foot-and-mouth disease outbreaks of the late 1960s. The present view of the Trust is that examples of the Blue Albion today have been 'reconstructed', probably from combinations of Friesian, Shorthorn and Welsh Black, and that in any event the breed was originally developed from Welsh Black/Shorthorn crosses well into the 20th century. Essentially, it was not possible to *prove* continuous existence for 75 years (the Trust's present criterion): the current stock cannot, on paper, be traced back to the original stock. It is the same for several breeds: they suffered from a lack of proper registration or, in some cases, the benefits of a proper breed society and herdbook or flockbook.

The long-running dispute about the Oxford Sandy and Black pig was along similar lines: that there seemed to have been a break in the continuity of breeding in the 1960s and that it was possible to construct a reasonably similar type by crossbreeding. Even before the Trust was formed, there had been increasing doubt that any pure specimens of the breed existed. In the summer of 1969 there was a rash of correspondence between Ray Carter at Stoneleigh, Arthur Manchester (at that stage Deputy Secretary of the NPBA), Ann Wheatley-Hubbard (who had known the genuine breed well in her youth) and others. Mrs Wheatley-Hubbard was approached by Carter in July for her advice: the admittedly crossbred Oxford Sandy and Blacks at Stoneleigh were causing problems; they didn't really know what to do with the animals and she was asked to help them find out more about the breed.

Back in 1964, in response to Rowlands's article in *Nature*, Philip Ryder-Davies had informed ZSL that only three people were keeping the breed in 1963 and that only two of those remained by 1964: Mr E.V. Holloway of Shothampton Farm, Charlbury, Oxon, had one young boar and five sows; Mrs V. Boseley at Sarsden Lodge, Churchill, Oxon, had five sows and no adult boar, though one sow had a young litter which included a male. Rowlands now passed the correspondence to Carter, who followed up the two leads and in October 1969 was informed by Mrs Boseley that she was the last in England to breed them 'but it meant inbreeding as no one else had a boar but I have regretted it ever since ... I am told the only way to start them again is a Large Black sow with a Tamworth boar.' Another letter, from Mrs Blackwell at Dean, near Charlbury in Oxfordshire, addressed to Mrs Wheatley-Hubbard, said: 'We are now itching to know what the Sandy we put to the Berkshire will produce, keeping our fingers crossed for a nicely marked boar!!! The young spotted gilt has grown well (the one we bred from the Berkshire gilt cross Tamworth).' Clearly there was a great deal of crossbreeding going on in 1969, and had been for several years. Holloway was dead by then and his son had given up pigs when 'the last OSB boar died'. The arguments would continue for another two decades.

Then there was the saga of the Glamorgan cattle, a breed closely related to the finch-backed Gloucester (whose own purity was challenged by some). In 1979, *Farmers Weekly* carried an article entitled 'Lost breed lives on in a corner of Sussex'. It referred to a group of cattle belonging to one Major Savage, by then in his eighties. When Richard Cooper took a look at them, they seemed to be a mixed group of Longhorns and Gloucesters running together which sometimes produced calves with the old Glamorgan colour pattern. Cooper was informed by Savage's agent that in fact they were three herds of Longhorn, Gloucester and Glamorgan cattle established respectively in 1884, 1834 and 1820 with cattle bought from Highgrove, Tetbury, in Gloucestershire. In an advertisement for the sale of the cattle in March 1979, the description was: 'Consignment of ten to twelve Glamorgan cows and heifers etc. The cows have been running with a Glamorgan bull.' Alderson subsequently published a letter in *Farmers Weekly* pointing out that 'neither the article in the *Farmers Weekly* nor any information that we have been able to obtain has substantiated the claims made for these cattle.'

Among the sheep breeds, the 'badger-faced' Welsh sheep known as Torddu and Torwen were rejected for a different reason: they were thought to be merely colour varieties of the Welsh Mountain. Their names, respectively, mean 'black belly' and 'white belly': they are, so to speak, negative/positive images of each other in their colouring. However, the Balwen has been accepted by the Trust, having justified its credentials as a distinct and separate breed.

There are one or two other breeds which some say are no more than colour variants (for example, Belted Welsh and White Galloway cattle) and the question of when a distinctive colour or pattern is adequate to define a variety as a breed in its own right is frequently debated. What *is* a breed?

Next was the goat problem. There have only ever been two goat breeds in the Trust's priority list: the Bagot and the Golden Guernsey, and even they are not free from controversy. But the Trust baulked at including the Old English goat, which it deemed to have been created simply by finding goats of a certain colour pattern (regardless of other characteristics or of ancestry) and calling them a breed because of that pattern. Again, what is a breed? Some of the examples just given smack of Jim Hindson's famous Three Black Spots syndrome.

A little less contentious, but by no means instantly accepted by the Trust, were the Eriskay pony and Castlemilk Moorit sheep. There was the special case of the remnant Norfolk Horn, part of the original Whipsnade gene bank and possibly the major spur to the formation of the Trust in the first place, which had been saved (just) by heavy back-crossing. During its 'rehabilitation' it was temporarily given the name New Norfolk Horn. All these were breeds which had to *prove* their credentials, rather than breeds which received automatic acceptance as their numbers fell.

The Trust was not much interested in horses in the very early days. Perhaps there was a general feeling that horses were not farm livestock — or that only the heavy horses counted in that category, and these tended to have their own very active and quite jealously protective breed societies. Again, it tended to boil down to personalities: if a Trust member felt strongly enough about a horse or pony breed, they would push its cause.

Ferals

Quite a different problem was whether the Trust should extend its support to feral animals, especially feral goats and sheep. The great champions in the feral cause, over many years, were Juliet Clutton-Brock and her husband Peter Jewell, who argue that the ferals harbour a potentially valuable pool of genes. The couple found themselves arguing in vain. Apart from a little support for Chillingham cattle, the Trust firmly fended off ferals, though in 1977 it had written to the New Zealand government in support of Betty Rowe, an American living in New Zealand, who was running a campaign to save the feral island sheep of Arapawa which local ecologists and agriculturalists wanted to eradicate. The Trust's support did have an effect and Mrs Rowe, heartened, continues to rescue various New Zealand feral populations today.

Jewell and Clutton-Brock, among others, persisted in pleading the feral cause in the UK for several more years. Then in 1989 the Trust organised an international conference at Warwick University and suddenly their persistence finally paid off. As a result of discussions at the conference and a meeting with representatives of the Nature Conservancy Council, the Forestry Commission and the Macaulay Land Use Research Institute at Penicuik in Lothian, the Trust changed its general attitude — not to such an extent that it embraced ferals warmly but at least it took a first tentative step or two and began to show just a little interest in them: it added a new Feral category to its priority list. Clutton-Brock, who has written some important books on the domestication of animals, is one of the few who has attempted to answer the question: 'What is a breed?'

Poultry

The acceptance criteria applied to poultry are slightly different from those for other rare breeds. For its first two decades the Trust did not include poultry at all, with the single exception of its support for the importation (to Andrew Sheppy's Cobthorn Farms near Bristol) of a breeding group of Scots Dumpies from The Hon Mrs Violet Carnegie in Kenya. In the days of the ZSL's gene bank a decision had been taken to disperse Whipsnade's collection of rare poultry breeds, largely to the Duchess of Devonshire. From time to time the possibility of bringing poultry into the Trust was raised by individuals, as several members were deeply involved in the breeding of rare poultry and waterfowl. For example, Christopher Marler had a fine collection at his Flamingo Gardens and Zoo Park at Weston Underwood and was a leading figure in the British Waterfowl Association (the secretary of which for several years happened to be Alderson); Geoffrey Cloke had an ancestral flock of Croad Langshan and high-performance White Sussex; Andrew Sheppy owned a large collection of breeds and had been secretary of the Rare Breeds (Poultry) Society.

In April 1979 Alderson wrote to Cloke, Marler and Sheppy seeking their views on including poultry in the Trust's programme. There were several voices in favour of the idea but some were worried: the 'fanciers' in the poultry world bred for exhibition and went for looks rather than pedigree — they were not averse to adding a bit of this and a bit of that to achieve their champions, and that was sacrilege to the Trust. It would be very difficult to ensure that a breed was pure (there was a distinct lack of record-keeping and registration) or met the criterion of continuous existence as a breed over a long period. And there were so many, many poultry 'breeds' that the Trust could find itself overwhelmed by them. The idea of including poultry was therefore rejected once again.

However, the Warwick conference in 1989 changed their minds. Roy Crawford had given a paper on poultry conservation which convinced the Trust that it could include poultry in its remit without jeopardising its principles if it concentrated on the themes of traditional British breeds (with long histories), utility characteristics, welfare, environment, and a high quality product. There were only about ten or a dozen British breeds which could meet these criteria.

The present situation

The basic criteria of the acceptance procedure (Appendix I) apply even now: the text of the original form drawn up in September 1976 has altered very little since it was first drafted. In a deliberate attempt to exclude 'reconstructed' breeds, item A4 has been changed so that a breed must have existed continuously (rather than simply have been known) for 75 years.

Today the Trust accepts breeds if, broadly, the numbers of breeding females fall below 1,500 for sheep, 1,000 for horses and ponies, 750 for cattle and 500 for goats and pigs (the latter has been increased from the old level of only 150). That is very low in comparison with the FAO's levels of something like 10,000 for breeds in developing countries and in some Mediterranean countries.

The titles of the priority categories have changed over the years and now agree more closely with those of the IUCN. They are: Critical, Endangered, Vulnerable, At Risk, Imported and Feral, and have been extended recently so that the Trust can include Minority breeds (those which have been on the list but have become successful enough to move off it) and a 'Watch' section for breeds moving the other way — they have never been on the list but their population is in decline and should be monitored. By way of contrast, the original breeds in their different categories in 1976 are listed here along with those in 1983 and 1993 (Table 4.1). It is interesting to note the movements that have taken place in the meantime as the relative fortunes of the breeds change over the years.

In 1994 there were major revisions, with greater use of population size and founder numbers as the essential criteria. Several other breeds fell within the new definition, including Lincoln Red and Devon cattle, and Llanwenog, South Wales Mountain, Kerry Hill, Devon and Cornwall Longwool, Whitefaced Dartmoor and Dorset Down sheep.

Table 4.1: The Priority Lists

Category	1975	1983	1993
SHEEP			
I	Portland Manx Loghtan Leicester Soay Wensleydale	Portland Leicester	< Castlemilk Moorit < Norfolk Horn
II	North Ronaldsay Cotswold Wiltshire Horn Lincoln Shropshire Shetland Oxford Down	Manx Loghtan Wensleydale North Ronaldsay Cotswold Lincoln	Whitefaced Woodland Portland North Ronaldsay < Balwen Leicester
III	Whitefaced Woodland Hebridean Ryeland > Southdown > Teeswater	Soay Whitefaced Woodland Hebridean Shropshire > Oxford Down	Manx Loghtan Soay Wensleydale Cotswold Hebridean < Hill Radnor
IV		Wiltshire Horn > Shetland > Ryeland	Wiltshire Horn Lincoln Shropshire < Southdown < Greyfaced Dartmoor
GOATS			
I		Bagot	Bagot
II	Bagot Golden Guernsey	Golden Guernsey	
IV			Golden Guernsey
CATTLE			
I	Irish Moiled Shetland White Park Gloucester	Irish Moiled Shetland White Park Kerry Gloucester	< Vaynol Irish Moiled Shetland
II	British White Longhorn	British White	White Park Kerry
III	Kerry > Northern Dairy Shorthorn Dexter	> Longhorn > Dexter < > Belted Galloway < Red Poll	Gloucester < Beef Shorthorn Red Poll
IV			British White
PIGS			
I	Tamworth Middle White Berkshire	Tamworth Middle White Berkshire British Lop Large Black	Tamworth Middle White British Lop Large Black
II	British Lop Gloucester OS	Gloucester OS	Berkshire Gloucester OS British Saddleback
III	Large Black	British Saddleback	
IV	British Saddleback		
HORSES			
I	Exmoor Pony	Exmoor Pony	Exmoor Pony Suffolk Cleveland Bay
II	Suffolk	Suffolk Dales Pony	
III	Dales Pony Clydesdale Fell Pony > Shire	Clydesdale > Fell Pony	Dales Pony
IV		< Cleveland Bay	Clydesdale < Irish Draught

Notes: < = entered Lists, > = left Lists

WENSLEYDALE

In its determination to help long-standing breeds which, by the vagaries of agricultural fashions, had slid into decline, the Trust needed a system by which individual animals had proper, central, supervised breeding records or pedigrees. However faithfully some breeders might keep records of their own stock, many others do not, and an animal with no pedigree has no way of proving that it is truly one of its breed — or did not, until recent advances in genetic fingerprinting, that is. But the coded secrets of DNA had not been unlocked in the early days of the Trust.

The bulk of the inherited Whipsnade gene bank was sheep, and very few of them had individual breed societies at that stage. One of the main functions of a breed society is to register animals which it accepts, by various criteria, as being true representatives of the breed, as well as promoting the breed as a whole by whatever means at its disposal. Now, that sounded like a useful role for the Trust. In cases where there was no breed society, perhaps it was logical that the Trust itself should become, in effect, a rare breeds society. But it didn't actually happen like that.

Combined Flock Book

In May 1974 Michael Rosenberg and Lawrence Alderson — acting privately through Alderson's company, Countrywide Livestock Ltd — published the first volume of a journal they aptly entitled *The Ark*. They also published, in the same way, the first volume of the 'Combined Flock Book of the Ark', which included: Cotswold, Manx Loghtan, Moorit Shetland (later Castlemilk Moorit), New Norfolk Horn, North Ronaldsay, Portland, St Kilda (later Hebridean), Shetland, Soay, and Whitefaced Woodland.

At the time, Rosenberg was involved in setting up a breeding programme for his own multibreed sheep unit at Ash Farm, Winkleigh, in Devon, in consultation with Alderson, but there was very little in writing about the histories and relationships of the gene bank and other rare breeds of sheep to give the programme a sound genetic basis. The Combined Flock Book was a private enterprise, completely independent of the Trust, and they were therefore able to include in it any breeds they chose, regardless of the Trust's priority lists. Thus they eventually managed to include the 'New' Norfolk Horn (the result of the breeding-back programme in which Alderson was involved as an independent consultant) and what they initially called the Moorit

Shetland (later Castlemilk Moorit).

The first volume was distributed free of charge to the 65 breeders of registered stock and to 45 subscribers; other copies were sold. The Trust was informed of its publication and simply accepted that it had been done, and done privately. That was how it remained until 1980, when the two men handed the Combined Flock Book over to the Trust for future management, at which stage the Trust was morally obliged to accept any animal already registered. Alderson handed his chairmanship of their Registration and Inspection committee to Ken Briggs.

When the project was passed to the Trust, a conscious decision was taken to attempt to discover and include in the programme the significant number of unregistered sheep that existed. The success of this policy was due largely to Cathie Church, who was rigorous in her pursuit of registrations, and Frank Bailey. Bailey was asked by Rosenberg to search out groups of unregistered stock and he travelled widely and successfully on this mission in the south of England. He made some surprising discoveries. On one farm, three Portland sheep obtained from a major flock in Kent were presented for inspection: one was pure black!

At an early stage, Rosenberg and Alderson had formed a Registration and Inspection committee to set standards, to help breeders who were perhaps inexperienced and to inspect every flock before it was included in the Combined Flock Book. The first meeting of this committee was held in the NAC conference hall on 20 January 1976, with Alderson in the chair; its secretary was Alastair Dymond, who at that stage was employed by the RASE, not the Trust.

The cast list at the first meeting of inspectors included, in alphabetical order, Ken and Nancy Briggs, Derek R. Dunston, Mr and Mrs G.B. Grant, L. Hindmarch, John H. Latimer, Wally W. McCurdie (the NAC head shepherd), Mrs B.D. Platt, D.R. Randall, Michael M. Rosenberg, Denis S. Vernon and J.S. Wood. This group of flock inspectors was an ambitious one: they intended to cover the entire country between them, though they immediately had to find an extra inspector to cover Scotland. They divided the map into regions and chose central towns in each so that inspectors and local groups of breeders could gather conveniently (and informally) at intervals.

The meeting resolved that absolute identification of each animal was essential, preferably by means of tags, or by tattooing or ear-notching; that in exceptional circumstances flocks could enter a 'supplementary flock registration' but all rams must be registered independently and no rams could be sold from supplementary flock ewes unless both parents were identifiable.

It was noted that the basis of the first volume of the Combined Flock Book (1974) included flocks which had not been inspected; there had not been the resources to do so. Such registration could continue but inspectors were asked to take advantage of inspection between lambing and weaning to recommend whether the stock was fit to be registered. If the animals showed any doubtful features, registration could be refused. The identification of parent stock was important, especially with animals which were being graded up. The aim was to complete all flock inspections well in advance of the Show & Sale in order to ensure that all entries were valid and identifiable.

As well as inspection and registration, the inspectors would encourage breeders in the good management practice of recording: they would supply forms for lambing records, weighing sheets and annual returns and would encourage the adoption of flock prefixes and also names for individual animals.

There was quite a discussion about grading-up (the practice of continued crossing to a purebred animal in order to increase the proportion of the latter breed's blood in a group) but they eventually decided in favour of the gradual closure of the Flock Book to graded-up animals. Breeds with higher populations should not be eligible for grading-up registration. Only registered rams could be used in grading-up programmes, and present grade stock would only be eligible for full registration after four top crosses and an inspection.

Then they talked at length about breed specifications — one of their biggest problems. Some of those early inspectors admit now that they really had little idea what they were looking at. It was early days, and the Trust was still on a strongly upward learning curve.

In 1978, a letter from Rosenberg, signing himself as vice chairman of the Trust, was typed on *The Ark* headed paper from his Winkleigh address, beneath which was the statement 'Registration facilities for ...' followed by a list of sheep breeds included in the Combined Flock Book. Clearly *The Ark* and the Flock Book were run together, and privately, from Winkleigh. The Flock Book itself stated that it was published by Countrywide Livestock Ltd. Even today, more than a decade after the Trust took over the Flock Book, sheep registered in it are given the ear-tag code CL — the initials of Countrywide Livestock.

Soon after the Trust took it over, the tight system of inspection began to drift because of lack of resources. In March 1981 Alderson produced a report making a plea that at least the important categories (new flocks, provisional registrations, grading-up and Category 1 breeds) should still be inspected fully. In fact the Trust eventually dropped flock inspections entirely and relied on breeders to be both knowledgeable and honest in registering their animals. It was quite possible for someone to register an animal which nobody else had ever seen and, although the registrant was obliged to sign a confirmation that the stated details were correct and that the animal conformed to the standards of the breed, some fairly strange ones turned up now and then. Alderson recalls a Portland ram at one Show & Sale which looked more like a North Ronaldsay.

Alderson's 1981 report had been printed on headed paper which bore the red banner 'Combined Flock Book', giving the address (still) as Countrywide Livestock Ltd, Market Place, Haltwhistle, Northumberland, and stating that it provided registration facilities for 'Castlemilk Shetland; Cotswold; Hebridean; Manx Loghtan; New Norfolk Horn; North Ronaldsay; Portland; Shetland; Soay; Whitefaced Woodland'. The Hebridean had originally been called the St Kilda and its tag code remained as the letter K.

He pointed out that Combined Flock Book sheep had been inspected twice: on the farm and again at the Show & Sale. The farm inspection covered four categories: new flocks, Provisional Register animals, grading-up animals, and breeds in Category 1 of the Trust's priority lists. The aim was to 'ensure that only animals of pure breeding are brought into the registration programme, and that grading-up animals conform to the standards of the breed'. At that stage the grading-up register admitted only seven-eighths bred animals.

The Show & Sale inspection ensured that animals shown or offered for sale were (a) registered (a condition of entry); (b) typical of the breed; (c) disease-free; (d) free from defects and deformities; and (e) in good condition. 'The basic criteria of these inspections is that animals must be fit for breeding within the registration programme, and must support the image of the Trust as a body with a serious and constructive interest in animal breeding and welfare. Defective and emaciated animals would defeat this objective.' This was a strong plea for professionalism in what was still sometimes the dilettante world of rare breeds.

The report also suggested that animals presented at the Show & Sale which were not in fact typical specimens of the breed should be down-graded to the provisional register or, in some cases, removed entirely from the programme. It would be another ten years before this idea re-appeared in the guise of the 'white' card under the Card Grading system.

Naturally there were many little arguments about which sheep could be registered in the CFB. As Technical Consultant, Alderson was often asked to advise the CFB Panel which had taken over from him in 1980 and which became answerable to the Breed Liaison committee. For example, there were some moorit Hebrideans: were they genuine Hebrideans or had the red gene come through faulty breeding? Then there was the question of grading-up programmes, and on this Alderson admits to changing his early view that this was an acceptable method of increasing population size (such an important factor in keeping rare breeds viable). He later believed that there was too great a risk of inappropriate characteristics being introduced, and grading-up was eventually stopped, though there was a little more leeway for the most critically low-numbered breeds.

In 1987 the Trust introduced a computer system and the CFB registrations became part of it. The program continues to be developed and can now provide detailed information not only on pedigree and ownership but also on coefficient of inbreeding and founder ancestor effect.

As the Trust matured and came to terms with its general role, the populations of the breeds began to increase. In some cases they did so with such success that they were no longer eligible for entry in the CFB and their breeders began to form their own breed societies. The Cotswold Sheep Society, which existed before the Trust was formed in 1973, remained in the Combined Flock Book for some time. The Shetland breeders formed a group within the Trust structure long after the population had exceeded the numerical criteria but eventually they formed an independent Shetland Sheep Breeders Group. At the time of writing, the Hebridean breeders are well on their way to following suit: their breed's population is increasing rapidly but on a well organised and balanced basis under the watchful eye of David Braithwaite and Eric Medway, in close consultation with the Trust.

Breeders' groups had been formed for most of the breeds by the early 1990s. Manx Loghtan and Whitefaced Woodland sheep and Bagot goats had active groups which published newsletters and organised workshops. In 1991 the Portland breeders, who for many years had conducted a valuable and enjoyable programme without an official group structure, decided to follow the example of the others. It remains to be seen whether a formal constitution is compatible with their memorable social events of the past: the dinners at Weymouth and Cirencester in 1980 and 1983 and the sparkling 'picnic' gathering for the 1987 Open Day at Nallers Farm (by courtesy of Norman and Michelle Jones). Soay breeders initiated discussion regarding the formation of a group in 1993, leaving only the Norfolk Horn breeders as champions of the relaxed, informal approach.

Pigs

In 1986 the Trust set up a procedure for the registration of rare breeds of pigs in conjunction with what was then called the National Pig Breeders Association (NPBA). Geoffrey Cloke was a vital link: he had been a member of NPBA since 1945 and so it did not prove too difficult to gain the co-operation of the association. In practice registrations were sent via the Trust, which in effect subsidised breeders' membership of the association (now the British Pig Association).

This was a major accomplishment. Breeders who had been discouraged from registering pigs of rare breeds because of the cost were able to take advantage of the opportunity to register through the Trust at a reduced charge and it had a dramatic effect on registrations.

Shetland cattle

There is a unique registration scheme through the Trust for the small black-and-white cattle of the Shetland Islands. This breed has an interesting recent history. During the early 1960s the senior livestock officer for DAFS, James Barr, visited the islands and discovered no more than 35 pedigree females and only two pedigree males of this old crofters' breed, with another half a dozen on the mainland. He rescued the breed almost single-handedly, establishing a DAFS herd at Knochnagael, near Inverness.

In October 1973 the Trust's Technical Consultant heard that a senior Shetland breeder, Tommy Frazer, had some cattle for sale. He therefore made contact with him but they had already been sold. Alderson then contacted Hugh Bowie, secretary of the breed society, who estimated that there were 100 cows and one bull on the islands in April 1975. In October that year, Alderson visited the DAFS herd in Inverness and later took five heifers and a bull to establish three herds in England. At the same time a Mr Hunt of Banbury bought some direct from the Islands. In 1976 Alderson brought down another eight Inverness heifers and established two more herds. The following year he visited Bowie and Frazer and then talked to the Shetland Islands Council about the possibility of funding the breed but was unsuccessful. He brought a further three heifers and another bull down from Inverness to establish another English herd.

The mainland cattle were moved around in England and Scotland to a limited extent and one or two new units were established. The breed was also included in ABRO's multibreed trials. Early in 1980 Alderson was concerned about the breed and corresponded with Barr about the possibility of taking semen from six Shetland bulls. He had inspected old volumes (1912—1922) of the Herd Book, which were lodged in the Cambridge University library. On behalf of the Trust, Marshall Watson visited the Shetland Islands in the same year and tracked down 31 females and one bull calf — that was all.

Shetland breeders are independent by nature and no doubt there was some resentment at well-meaning interference from a bunch of people based in the English midlands. The breed society was not very happy either and a situation of some distrust rumbled on for several years, though the Trust did manage a useful piece of media coverage when it sent a Shetland bull and a group of heifers to help restock the Falkland Islands after the end of hostilities there. In 1987 Alderson and Cloke met the breed society officers in the Shetland Islands and established a strong personal link. Meanwhile in 1982 the Shetland Islands Council had been persuaded to deposit money with the Trust so that it could pay incentives to islanders to breed their cattle pure.

There remains a problem with the Herd Book. The local system prevents registration of any animal until it has itself produced progeny — and any breeder of livestock knows how difficult it can be to keep track of individual animals when they are young. In 1991 the Trust set up a system of birth notification: the birth of an animal is certified by the Trust, and the certificate accompanies the animal until it can be registered formally in the Herd Book.

CHAPTER FIVE
TECHNICAL PROJECTS
AND
MARKETING

GLOUCESTER OLD SPOTS

WHITEFACED WOODLAND

CHAPTER 5:
TECHNICAL PROJECTS AND MARKETING

The actual technical projects undertaken or supported by the Trust form its essential work on behalf of rare breeds. Those carried out during the past two decades divide broadly into three main and sequential phases: first, the saving of rare breeds by immediate emergency action in the early days, followed by conservation projects to consolidate 'saved' breeds; second, the evaluation of those breeds to discover their characteristics and qualities; and, finally, the utilisation and promotion of evaluated breeds by finding niches in which their qualities can best be exploited in order to increase their popularity, whether for non-commercial roles or by imaginative marketing.

THE PROJECTS PROGRAMMES

There are so many technical projects that sometimes it can be confusing. In the introduction to its 1979 projects appeal, when the Trust was only six years old, it was pointed out:

> Conservation by itself is not enough. Conservation of the 44 breeds with which the Trust is currently concerned can be achieved only by the willing participation and co-operation of the many breeders and farmers who have these animals. Some of the breeds which we believe must be kept because of their unique genetic qualities are not economically viable in their own right in today's market situation. The farming community has been hit, as everyone else, by inflation and is looking more to the short term when every penny counts rather than to possible long term benefit for future generations. This trend is worrying and even breeds which are secure now may shortly be at risk. There can be few who have not heard of and would not recognise one of the most attractive cows to grace the British Isles — the Jersey, but only recently it has been announced that the famous 'Ovaltine Herd' is to go under the hammer because it is no longer viable, bearing in mind its milk yield, the prices paid for high-quality milk and the effects of Common Market agricultural policies. If this can happen to a large famous herd of not yet rare animals ...

How often today are Jersey cows seen in the fields? Prevention is better than cure — a proactive approach is more effective than mere reaction to crises. Seeking the security of continuity for once, the Trust appealed for at least £125,000 over the next five years, and it carefully costed each of eleven proposed projects over that period. It also looked to the more distant future in view of technological developments such as the storage of frozen embryos and tissue typing. Genetic engineering was not yet in the crystal ball, but eggs and embryos have long played their part in the Trust's gene bank work.

That paper also gave a chronology of the Trust's development to date; it outlined the Rare Breeds Acceptance Procedure and listed the breeds by category as at November, 1978; and it made a strong case for the conservation of rare breeds — for agriculture, for archaeological and historical study, for genetic research, for a better understanding of environmental management in support of wild as well as domesticated animals, and finally for the sake of pure sentiment and tradition, with the emotive sign-off: 'A stuffed dodo does not compensate for the living and walking bird.'

The 1979 appeal listed several areas of interest: the survey; breed incentives; the Linga Holm project; congenital defects; analysis of breed structure; blood- and milk-

typing; chromosome analysis; characteristics of North Ronaldsay sheep; bull performance testing; evaluation of special qualities; polyunsaturated fats; ease of parturition.

The policy document prepared in 1991 included a list of projects divided into
categories: those that had been concluded, those that were ongoing, and those that
were planned for the near future. Subject to annual reviews by Council, this list forms
the basis for the Trust's programme of work.

SURVEYS AND REGISTRATION

The initial step in the protection of any rare breed needs to be a population survey. The
first of the Trust's surveys, carried out in 1974, identified many of the breeds that
needed the Trust's attention; the establishment of a Rare Breeds Acceptance Procedure created a method of screening the claims of additional breeds.

The natural extension of the initial surveys was the creation of registration
programmes where necessary — hence the Combined Flock Book in the early days
and more recent co-operation with the British Pig Association to facilitate the
registration of rare breeds of pigs. In later years the Trust has created a registration
programme for Vaynol cattle.

The final phase in any registration project is the publication of herd books or flock
books. The Combined Flock Book was published from its early days; the Vaynol Herd
Book is published; and the Trust has also provided registration facilities for breeds
such as Shetland cattle and produced a Herd Book on their behalf. The series of
surveys, registration programme and publication is seen as one integral project.

FOOT-AND-MOUTH DISEASE

The main emphasis of the Trust's work in its first three or four years had been on
emergency and conservation activities, such as those already described. Another
emergency project was the action taken during outbreaks of foot-and-mouth disease.
It was realised that some breeds were vulnerable in that they were very small
populations, perhaps bunched together in only one or two groups which could all too
easily be wiped out by the disease (or even by the Ministry of Agriculture in one of its
disease eradication schemes). In fact even before the Trust was formed Roger
Ewbank, then of the University of Liverpool, had written to Captain H.W. Dawes,
veterinary adviser to the RASE, pointing out this potential danger to the 'collection of
rare and unique breeds' of livestock which was then being gathered together at
Stoneleigh during the dispersal of the Whipsnade gene bank. Unaware of this
correspondence, the Trust itself would later debate the problem of disease hitting
animals concentrated in one place. Its 'Orkney project' (the purchase of Linga Holm
for a reserve flock of North Ronaldsay sheep) was conceived for that very reason; to
spread the risk, at a time when there was yet another epidemic of foot-and-mouth.

In the foot-and-mouth outbreak of March 1981 the breeds which gave particular
cause for concern were Portland sheep; Bagot goats; and British Lop, Gloucester Old
Spots and Middle White pigs. The critically endangered Irish Moiled cattle were
considered reasonably safe in their relative isolation in Ulster. The Trust got to work
quickly. Alderson was in contact with Ken Briggs, Geoffrey Cloke, Arthur Manchester, Michael Rosenberg, Ann Wheatley-Hubbard and Jim Hindson as soon as the
current outbreak looked serious, and within days they had arranged a positive musical
chairs: it was proposed to move a boar and two gilts of a rare line of British Lops from
Cloke's farm near Solihull, Warwickshire, to Appleby Castle in Cumbria (with a
supply of concentrates); Bagot goats from Briggs in Worcestershire would travel to
Mrs D. Miller in East Lothian (with a supply of hay); and six yearling White Park
heifers from Lennoxlove in Scotland would move north to Nairn.

CRYOPRESERVATION

The foot-and-mouth exercise highlighted what had become clear at an early stage: that it was necessary to undertake the cryopreservation of genetic material for long-term storage as an insurance against disaster befalling the live population. Thus the *Semen Bank* was created, which also enabled the development of an *artificial insemination* service to overcome the unwillingness, or inability, of many rare breeds owners to keep a bull or a boar. The Semen Bank is one of the most important projects undertaken by the Trust. It has been developed mainly around cattle breeds but pigs have been included more recently and it has been agreed, in principle, that it can be extended to sheep, goats and horses if necessary.

Embryo collection and transfer has been a more controversial subject and opinions have been divided between those who consider it to be the ultimate form of insurance and those who consider it to be a poor risk in terms of cost and benefit. Some successful embryo transfer work has been carried out privately with White Park and Irish Moiled cattle, and the Trust's main involvement has been with more difficult breeds. Two projects with Vaynol and Gloucester cattle were failures, though it seems likely that the latter may have been simply a problem of ignorance about the timing of ovulation in Gloucester cows.

BREEDING PROGRAMMES

The development of the Semen Bank enabled more effective breeding programmes to be carried out. These were often specific to a given breed but in general they revolved around the concept of *rotational mating programmes* devised by Alderson in 1974. These were used in several breeds, including Portland and Norfolk Horn sheep.

In the early years the main purpose of breeding programmes was to minimise the level of inbreeding. However, it became increasingly obvious that this was not necessarily a sensible criterion to apply in conservation programmes and in 1991 Alderson, in consultation with Rex Walters and Ian Gill (the Trust's genetic advisers), introduced the use of *effective founder number* as a more relevant measure. This is based on the contribution of the founder ancestors to an animal or to a breed and is thus a better measure of the likely conservation of the original genome.

The Trust is developing a sophisticated computer programme which calculates the *coefficient of inbreeding* for each animal as well as its effective founder number. In addition, it gives details of the founder animals and dominant ancestors from which those measures were obtained. This will prove of enormous value in the Trust's conservation work.

ANIMAL HEALTH

The programmes to save rare breeds are frequently hindered by health problems. The Trust has therefore placed considerable emphasis on the development of disease control and health promotion systems.

A register of *congenital defects* was established at an early stage. It was kept in a confidential file by Alderson and has provided valuable pointers to the causes of various diseases and defects. In particular, studies have been made of the occurrence of atresia ani in Whitefaced Woodland sheep and of entropion in Cotswold sheep (both the result of inbreeding), split eyelid in multihorned sheep, achondroplasia in Dexter cattle, and dystocia in cattle and sheep.

Such projects have demonstrated, for example, that split eyelid has never been found in two-horned sheep, and that the rare breeds which have bypassed intensive selection procedures suffer far fewer parturition difficulties than other breeds. On the other hand, they do suffer from low conception rates with artificial insemination and with embryo transfer. It is likely that the normal procedures used in these two processes are based on the physiology of modern popular breeds and that some rare breeds may not conform to those standards. The Trust became concerned by the waste of valuable genetic material as a result of low conception rates and organised a seminar to educate owners in the art of heat detection; it also continues to investigate the possibilities of breed variations in reproduction in rare breeds.

The imposition of *health standards* throughout the EC increasingly affects rare breeds and the Trust has taken the initiative by encouraging owners to participate in health programmes wherever possible. It has extended this policy by including classes for both accredited and non-accredited stock of any breed at the 1993 Show & Sale. There has already been a growing trend towards EBL accreditation among rare breeds of cattle, but the interest of sheep breeders in maedi-visna accreditation is not so evident.

BREED INCENTIVES

In recent years, especially under the guidance of Geoffrey Cloke as chairman of the Breed Liaison committee, there has been a move towards a proactive policy within the Trust — taking the initiative by anticipating problems and giving help before it is requested.

The Trust had been conscious of the need to supply financial assistance as well as technical advice and moral support in order that rare breeds could be saved effectively. The breed incentives programme became a major part of the work of the Breed Liaison committee and developed into a *five-year support programme*. Pig breeds illustrate this progression well. In 1980 the Trust set up a system of boar incentives for Middle White pigs with the intention of encouraging breeders to retain boars and use them for pure breeding. The system quickly spread to other rare breeds but it was never fully effective and was replaced by sow/litter incentives, in which part of the support was paid on the registration of purebred progeny, thus ensuring that the programme was carried through into the next generation. But even this was not as effective as had been anticipated and it was therefore developed into five-year support programmes. Meanwhile other pig incentives had been applied, particularly the importation of Australian boars in 1976 and again in the early 1990s.

Five-year support programmes are agreed in detail between the Trust and breeders' groups, so that everybody knows they can depend on continuing projects. By the end of 1993, fourteen breeds were participating in such programmes: White Park, Red Poll, Shetland and Beef Shorthorn cattle; Bagot goats; Hebridean and Manx Loghtan sheep; Suffolk and Irish Draught horses; and Berkshire, Gloucester Old Spots, Tamworth, Middle White and British Lop pigs. The type of support varied widely from display and promotional material to scientific studies and commercial evaluation trials. The programmes have already yielded interesting and valuable information, such as the demonstration of lower cholesterol levels in the carcase fat of some 'unimproved' breeds, the identification of potential dams of Red Poll bulls through milk recording, and the evaluation of wool quality in Manx Loghtan sheep. There is no standard system of breed support: each programme is tailored to the particular needs of a breed. An early example was the creation of a reserve population of North Ronaldsay sheep on Linga Holm to increase the security of the breed.

There has also been a *poultry programme,* developed after 1989, designed to save not simply the traditional breeds but also their characteristics. This is potentially an enormous project: the intention is to create four accredited units, each of 50 hens and 15—20 cocks, for each of the ten or so breeds on the Trust's list. That might not sound dramatic, but the very narrow genetic base of the poultry used by international breeding companies, and the increasing importance of animal welfare and environmental considerations as well as the quality of the products, are likely to give this project a high profile as time goes by.

CO-OPERATION

In many cases the Trust co-operates in technical projects with other organisations. Such co-operation goes back to the Trust's origins from the joint meeting of interested parties convened by the Zoological Society of London. Reading University had accepted some of the original Whipsnade gene bank animals and also provided students to undertake the original breeds survey; later, it worked with the Trust to set up a computerised analysis of breed structures and a study of *genetic distance* in cattle breeds — the Trust funded a revealing doctoral thesis, 'Polymorphisms of rare breeds of cattle', by Nicola Royle, who calculated genetic distance and also the heterozygosity of various breeds. For example, she was able to show the great genetic distance between White Park and other breeds. She identified a distinctive C-band translocation in the White Park and also calculated the incidence of Robertsonian translocation (1/29) in the British White.

Biochemical research has proved invaluable. Since its early years, the Trust has supported blood-typing work, including a project at Newmarket with Exmoor ponies and another at Dublin University (under the supervision of Eamonn Kelly) with Irish Moiled and Kerry cattle. Dr E.M. (Betty) Tucker had been building up a substantial blood-typing library for rare breeds of sheep since the days of the Whipsnade gene bank. But blood-typing has now been overtaken by *genetic finger-printing* and the Trust is co-operating with the University of Liverpool, through Ian Gill, in a project based on Irish Moiled cattle. The Ministry of Agriculture has determined that characterisation of rare breeds, in order to demonstrate their unique qualities, is a prerequisite for any possible financial support from the state; the Chief Scientist's group has further decided that such characterisation should be based on DNA analysis. The Trust, while not wholly in agreement with this policy, is co-operating in the development of appropriate projects.

Other universities have become involved in the Trust's work: for example, London University and the study of *ease of parturition* in sheep; Cambridge's studies of *behaviour in North Ronaldsay sheep*; and Bristol's *genetics seminars*. The West of Scotland College of Agriculture has carried out work on the measurement of *cholesterol and polyunsaturated fats* in the carcases of Hebridean sheep.

On the agricultural industry side, the Milk Marketing Board has co-operated for many years with the *semen bank* — the Bank of Genetic Variability or, in the MMB's phrase, the 'museum' bank. The Meat & Livestock Commission has co-operated with *performance testing* for bulls of rare breeds. The national *livestock associations* are all involved in the Trust's work — the National Sheep Association, the National Cattle Breeders Association and the British Pig Association.

Then there are various *conservation and heritage* bodies with which the Trust has worked, and will increasingly do so in the future. The British Museum (Natural History) is a strong link through its archaeozoologist, Juliet Clutton-Brock: she was collecting rare-breed skeletons even before the Trust was set up, and several early Trust meetings were held on the Museum's premises. There are active links, either through the Trust's members and officials or direct, with the National Trust and with nature conservation interests such as the Countryside Commission, the Ministry of Defence and English Nature. The latter was formerly part of the Nature Conservancy, a link which precedes the formal creation of the Trust. In a category of its own, a very recent and increasingly involved organisation is *H.M. Prison Service*: twelve of its centres maintain breeding groups of rare breeds, and the Service co-operated with the Trust in the 1992 Hyde Park event.

OWNERSHIP OF LIVESTOCK

There are other ways in which rare breeds are being actively promoted and the Trust is continually identifying new opportunities to establish breeding units. In practice it is rare for the Trust itself to own animals: it simply does not have the resources or facilities to farm and breed them, though that is still a dream.

The Trust's ownership of livestock began with the *Linga Holm* flock. Its next group of animals was a very small flock of *Bagot goats* bequeathed by Nancy, Lady Bagot, in May 1979, by which time the flock at Blithfield Park had fallen to very low numbers. The system of 'ownership' by the Trust in this case developed into one which set a pattern for other breeds: the *agistment scheme* by which co-operating breeders maintain the animals on the Trust's behalf, initially in return for an agistment fee from the Trust but later at the breeders' own expense. The breeders expand their unit and in due course give back to the Trust the same number of animals as was originally placed with them (reminiscent of the African brideprice system in which a husband receives a number of animals from his father-in-law at the marriage but must return the same number if the marriage fails!). The breeders then have ownership of the remaining animals, while those that have been returned to the Trust are placed in other centres to repeat the process. The Bagots, for example, have been at Wimpole Home Farm near Cambridge, Graves Park in Sheffield, Layer Marney Towers in Essex, Tilgate Park in Sussex, and Sherwood Forest Farm Park in Nottinghamshire.

The Trust was also bequeathed a group of *Vaynol cattle*, an exciting venture as the animals had lived as ferals for decades: they were not used to being handled and did not appreciate being loaded, transported and removed to alien surroundings. The herd was registered in a section of the White Park herd book until 1987 but, like the 'wild' Chillingham cattle, they conformed neither to the herd book rules nor to the description of the White Park breed. They are now regarded as a separate breed. The history of the Vaynols has been traced back to Argyllshire in the early 19th century and includes, intriguingly, experimental crossing with Indian zebu (humped) cattle during the 1880s, traces of which persisted in some animals a hundred years later. There were other influences, including white Highland cattle.

The herd was transferred from Scotland to Vaynol Park in North Wales during the 1870s and 1880s, and ran at various stages with Chillingham x Cadzow and Dynevor bulls until the herd was closed in 1930. Fifty years later there was only one adult bull in the herd at Vaynol, with three yearling bulls, nine cows and three heifers. In 1980 the estate was sold and the entire herd was transferred to Shugborough Park Farm in Staffordshire, with financial assistance from the Science Museum. In 1984 the Friends of Shugborough Park Farm gifted the herd to the Trust and the animals were agisted initially to Henson's Bemborough Farm in Gloucestershire, with the Trust paying for the herd's keep and management. Part of the cost was met by the Livestock Support Fund created by a legacy from Mrs Flora E.A. Gregory. The following year, to avoid the potential vulnerability of having the entire herd on one site, a small group was transferred to Graves Park in Sheffield. In 1987 the remainder of the herd was moved to St David's, Dyfed, with one young bull going to Temple Newsam estate in Leeds. In 1989 the St David's group was also transferred to Leeds. In the same year, noticing that an increasing proportion of the calves were black (which would have been culled in the original Vaynol Park system), it was agreed to make no discrimination against colour and also to recognise the Vaynol as a separate breed.

The Vaynols have been the subject of several studies: temperament (unreliable!), udder problems, calf mortality, mineral deficiencies, reactions to tranquilliser drugs, embryo collection (unsuccessful so far), inbreeding levels (the herd had already experienced and emerged through a severe inbreeding bottleneck), semen collection nd so on. However, their fortunes changed with the move to Yorkshire and they were ositively thriving in the two centres. At the end of 1992 it was possible to send a group rom Leeds back to the breed's original home, Vaynol Park, still on the agistment cheme and still under the watchful eye of the Vaynol herd book committee's members: Geoffrey Cloke as chairman of the Trust's Breed Liaison committee, Lawrence Alderson as the Trust's Executive Director, Bernard Lewis as the manager of project developments for the City of Sheffield Metropolitan District, and John Tinker as Leeds City Council's director of parks. Tinker, a plain-talking, pipe-smoking Yorkshireman, began his working life in farming and one of the parks under his determined and imaginative control was the country's largest rare breeds farm, that at Temple Newsam.

There was another example, albeit indirect, of Trust ownership. When Michael Rosenberg ceased farming at Ash Farm, Winkleigh, he donated his Longhorn cattle to Wimpole Home Farm along with the White Park cattle he had kept on an estate in Northumberland, but his Shetland cattle and the various sheep breeds were given to the Trust for distribution as it saw fit. They were all dispersed on the agistment basis in the same way as the Bagot goats.

The Trust occasionally purchases *boars* when it needs to collect their semen or as an emergency measure. For example, the famous Playle herd of Large Blacks in East Anglia was within a day of being dispersed when the Trust heard about it. Viki Mills sprang into action and made sure that Geoffrey Cloke and Philip Snell attended the sale. They managed to buy the important bloodlines on the Trust's behalf and the animals were appropriately placed.

The Trust will also buy *bulls* where necessary in order to obtain semen for the semen bank, though in most cases bull owners are generous enough to allow semen collection.

The ownership of livestock, however, is not a high priority for the Trust. Its philosophy is that livestock should be kept by livestock keepers but that the Trust will offer financial incentives, advice and other forms of support as appropriate.

SHETLAND

TRUST FARMS

Over the years there has often been discussion about whether or not the Trust should acquire its own farm or farms for breeding rare breeds. The question arose, for example, when the National Trust offered Greys Court, a collection of farm buildings near Henley on Thames, Oxon. The Trust took a look at it and was tempted: it offered potential as a centre for holding bulls and boars, or for evaluating breeds or as emergency housing for stock, but it would have been hugely expensive in terms of capital investment and the commitment to day-to-day management. The idea came up again in 1977 when the Yorkshire Agricultural Society was prepared to make buildings, land and stockman's accommodation available to the Trust at the Harrogate showground: again, the commitment was considered too great, but Rosenberg and Alderson were authorised to negotiate on the basis that the Trust might provide educational material and displays and some supervisory management and expertise if the YAS (rather than the Trust) would set up such a centre.

It remained a dream which would not go away. From time to time Cloke, in particular, who envisaged a Trust stud for boars and bulls, would come across potential properties on the market but again and again the cost factors mitigated against it. The closest they came to it was probably in late 1986 when the MLC was pulling out of its centre at Selby. Cloke and Alderson went up and talked to Dr Hugh Reed about the possibility of buying it; they drew up budgets and balance sheets, they explored all the possibilities, but ... it remains a dream. A few months later someone offered the Trust a small farm with buildings at Lower Kingcombe in Dorset and, again, the problem of ongoing costs damped the spark. Yes, it remains a dream for Cloke: a stud farm, if only to hold animals and collect semen from them, and perhaps hire out the stock to breed the next generation, under strict control so that the Trust knows exactly what is happening rather than hoping that other people do. He would love to see the Trust's own stud farm but accepts that, realistically, it would never pay.

UTILISATION

Evaluation

Once a breed has been saved, there are two different approaches to the exploitation of its characteristics: the ecological angle and the marketing angle. But first of all those characteristics must be identified and then evaluated.

Evaluation need not be restricted to the absolute dependence on high production levels required by commercial farming — the numbers game of how many eggs a hen can lay, what sort of lambing percentages a ewe can reach, how many gallons of milk a cow can produce (regardless of the milk's actual quality). These criteria of efficiency in production have been heavily accentuated by advisory services for some twenty years or more, and all tend to rely on substantial inputs to gain substantial outputs. But that is not necessarily efficient. In recent years the balance has swung towards economy of output: that an animal makes best use of the inputs. Several studies have demonstrated the greater efficiency of meat production by rare breeds. Others are looking at milk quality by measuring output in weight of protein rather than weight of whole milk, or relating output to cost of feeding, or weighing up the economic advantages of ease of parturition.

The *North Ronaldsay sheep* is a typical example of the Trust's work in, first of all, saving a breed and putting it on a more secure footing, and then turning attention to determining the breed's characteristics and evaluating its qualities. This is a continuing process with the North Ronaldsay. Paterson's early behavioural study ('The foraging strategy of the seaweed-eating sheep') was published as a doctoral thesis from Cambridge University; it was followed by a parasitology study by David Britt from the

University of Liverpool and a programme to monitor the incidence of cryptorchids among the sheep on Linga Holm.

When the *poultry project* was started, it was discovered that several breeds had lost many of their utility characteristics. Before the accredited units could be developed, it was necessary to establish evaluation units and in some cases birds with very creditable performance levels were found. Geoffrey Cloke's Croad Langshan stock produced a hen-house average well in excess of 200 eggs, while foundation stocks of White Sussex from Cloke and Geoffrey Marsden both included birds which yielded in excess of 250 eggs. The Scots Grey has retained much of its production potential but, in contrast, there is considerable variation among breeds such as the Old English Pheasant Fowl and the Dorking, while the extreme emphasis on phenotype (i.e. what the bird looks like) in the Derbyshire Redcap has virtually destroyed that breed's utility characteristics.

Michael Rosenberg appreciated at an early stage that it was necessary to discover and identify the normal standards and characteristics of rare breeds. He therefore established at Ash Farm a detailed *recording system* which measured performance and also graded all animals for phenotypic characteristics. This system, implemented at Ash Farm over a period of 15 years, gave the Trust a ready-made foundation on which to continue its characterisation programmes for rare breeds of sheep. The later adoption of card grading was an extension of this programme and was based on the same philosophy.

The ecological angle

After evaluation, one way in which to exploit a breed's characteristics is to make best use of them in noncommercial situations and there are several examples of the way in which their hardiness, thriftiness and other traditional qualities have been put to practical if not financially rewarding advantage. The key word here is probably 'ecological' benefit. Primitive sheep breeds seem to be the stars in this respect: for example, Soay sheep have been used for grazing inaccessible or unproductive areas such as English China Clay's spoil tips in Cornwall. No-nonsense breeds like these do not need lush pasture or intensive management of any kind: they simply get on with living (for which they have far more of a liking than most sheep!) and making the most of what they can get.

The Hebrideans were found to have an important ecological niche of their own in their ability to control invasive scrub and weed species, as demonstrated in trials at Spurn Head and Skipwith Common (near York). The North Ronaldsay sheep have the unique ability to survive on an exclusive diet of seaweed. Several of the older rare breeds of sheep, and of cattle, which have not been subjected to intensive selection pressures demonstrate a preference for coarser, rougher grazing.

The marketing angle

The other approach is to look at rare breed products, assess what is special about them, and then get out there and sell those products, exploiting the whole idea of rare breeds at the same time. This, indirectly, is the essence of the farm parks but there is also the more direct commercial marketing of products — meat and wool in particular.

An early example of evaluation in this respect was a demonstration by the North of Scotland College of Agriculture that the productivity of *Shetland sheep* was more efficient than that of the Scottish Blackface. Subsequent trials by the Trust demonstrated a superiority for *Hebridean sheep*, not only in comparison with the Scottish Blackface but also the Mule in lowland lamb production systems. Annual returns made by breeders registering stock in the Combined Flock Book are analysed to identify performance characteristics in various rare breeds and to indicate possible further candidates for commercial evaluation in comparative trials.

A similar situation has been discovered in rare breeds of cattle. Private trials in the mid 1970s demonstrated the superiority of *White Park* bulls in beef crossing trials, in comparison with Hereford and Welsh Black. That superiority has been confirmed more recently in trials with the Limousin (currently the most popular beef bull in the UK): crossbred White Park calves outperformed their contemporaries sired by Limousin bulls. This probably derives from the significant hybrid vigour exhibited by crossbred White Park cattle as a result of their great genetic distance from other breeds, as calculated by Royle. Current trial work in cattle breeds also looks at milk production in order to ensure that the traditional qualities of dairy breeds are not lost. High-yielding *Red Poll* cows have already been identified during milk recording and the programme has been extended to a Kerry unit and to one or two Gloucester units, although many owners of Gloucester cattle are no longer interested in dairy production.

It is important that this process of evaluation and discovery extends beyond the assessment of phenotypic characteristics, and the Trust is conscious of the need to emphasise the high quality of rare breed products, especially meat. The anecdotal evidence for the special qualities of rare breed meat is often quoted — for example, in relation to Portland sheep and others — and in recent years it has been possible to substantiate these claims more precisely. A research project carried out by AFRC demonstrated that the *quality of the meat* in carcases of rare breeds of pig was significantly superior to that of popular commercial breeds such as Large White, Landrace, Hampshire, Duroc and (especially) Piétrain.

Another AFRC project on behalf of the Trust identified the proportion, distribution and location of fat in the carcases of four British breeds of sheep and demonstrated the *high lean content* in the Hebridean and Soay. Perhaps most important of all, preliminary results from a Trust research project have shown that carcases of Hebridean sheep, and probably of other 'unimproved' breeds of sheep and cattle, contain a lower level of *cholesterol* and a higher level of *polyunsaturated fats* than those of the popular commercial breeds. If these preliminary findings are confirmed, the implications for rare breeds and for human health are significant.

The star of rare-breed marketing must surely be Anne Petch, whose Heal Farm Meats is the major success story in the rare breeds marketing world. She comes from good stock herself. On her mother's side she traces back to prime minister Sir Robert Peel, who was also a collector of rare breeds. It is 'accepted family wisdom' that a dark red boar was brought back from the West Indies by sea to the family's Axford estate near Tamworth, Staffordshire, some time between 1810 and 1820. This Red Barbadian boar was bred with the estate's pigs and the red offspring were crossmated to produce a type known initially as the Axford pig. And that, some say, was the ancestor of today's ginger Tamworth.

When Anne married Richard Petch in 1971, she already had a couple of breeding sows, the first having been given to her as a sickly newborn. Later that year the couple acquired East Hele Farm in Devon. In 1973 she purchased a small flock of pedigree Dorset Horn sheep, some Devon cattle and an in-pig Gloucester Old Spots sow. The cattle in particular were a huge success and her first homebred bull became a champion, but she was increasingly fascinated by pigs. She found early success in showing her GOS and Saddlebacks but it was after she joined the Trust in 1974 that 'pigs got totally out of hand'. By 1977 she was keeping up to 60 sows of four breeds, with no more than part-time help.

Then the pig market went into one of its typical troughs and this dauntless woman reacted positively. She began to pilot 'Deep Freeze Pork' based on her far-from-intensively managed pigs: she had always stuck to the principles of high welfare and minimal use of drugs. She also spread her skills by running 'Whole Hog' weekend courses, to teach smallholders about pig-keeping from conception to consumption.

By 1979 she was ready for the major marketing stride of converting the farm's redundant dairy buildings into a butchery. After much experimentation and research into traditional recipes, she managed to establish a range of pigmeat products from her rare breeds, based on the old recipes and with the minimum of additives. Rare breeds of lamb were added to the range over the next four years, and Red Devon beef, both produced under contract by professional pedigree breeders.

In January 1981 her enterprise was featured on the women's page of the *Daily Telegraph*, and in the space of a fortnight she was contacted by an amazing 15,000 enquirers — people wanting to know more about her pure meat and also more about rare breeds of farm livestock. It took seven or eight months to catch up with all the orders, and all the processing (including curing and smoking the meats) was carried out at the farm butchery. Incidentally, although the farm name is East Hele, she added the professional touch to her marketing skills by calling the product range *Heal* Farm Meats.

Petch was the first of a new generation of producers who were determined to break free from the conventional commercial market and at first she was regarded as 'slightly odd'. But she had anticipated the public mood: people began to question the ethics of factory farming and the heavy use of chemicals in food processing, and Petch was there when they turned to her.

The excellent public relations continued. Heal Farm Meats began to appear in food guides and was featured regularly in magazines and newspapers. In 1984 Petch was runner-up in BBC Radio 4's 'Small Business of the Year' competition and won the NFU Marketing competition. The following year she won the Venture Cash Award for Innovative Research with her project on grass feeding and its effect of polyunsaturated fatty acid deposits in pigs. Over the years she has been widely acknowledged as a specialist in the production of rare breed animals for meat and as an expert in the presentation and manufacture of high quality meats and meat products.

In spite of such a time-consuming enterprise on the farm, she found time to be on the Trust's Council and its Breed Liaison committee, and on the BPA's Minor Breeds committee and the Devon County Agricultural Association's Food Hall committee; she was joint chairman of Devon Harvest and vice chairman and founder member of Devon Fare; she was on the British Meat Committee, Bath & West and Southern Counties Agricultural Association. She was a Board member of the British Culinary Institute, a Panel member of the South West Agricultural Marketing Unit, a local TAB member of the Training & Enterprise Council, a member for the South West region of the MAFF Minister's Advisory Panel, and chairman of the Berkshire Breeders Club.

That was still not enough to absorb all her energies. She acts as an independent consultant on farm diversification projects. Her name appears on several judges' lists for individual pig breeds and on all minor breeds judges' lists, and all the breeds lists for the BPA; she has judged at many county shows, the Royal Show and Smithfield. As for her Trust involvement, she has organised pig management workshops and she carried out the first national survey of Large Blacks, which triggered the comprehensive annual series of surveys of all the rare and endangered pig breeds.

Her latest venture is a Food Village project in co-operation with local councils and the Rural Development Commission to develop a site providing premises for a number of speciality food producers, all to EC standards on a site which will be 'ecologically sound': it will generate its own power (from locally produced biomass and from gases generated by on-site waste); it will re-process all its own waste and dirty water.

She is also investigating the possibility of an on-site humane slaughter unit and is one of several who are concerned at the way in which recent EC legislation is closing down many local slaughterhouses so that animals will have to be transported over considerable distances, with all the welfare problems that can cause. (There are moves afoot to introduce mobile slaughterhouses — a scheme propounded by the Humane Slaughter Association.) As a sideline, Petch and Ken Briggs are considering the movement of North Ronaldsay sheep to see if they can be marketed in the south.

In her 'spare time' this mother-of-two flies a 25-year-old Piper Cherokee light aeroplane and manages a couple of trips a year to France for research into food production methods and to explore regional cuisines. In anybody's marketing book, all that is a very hard act to follow but it provides a strong incentive.

*

On a less ambitious note, but praiseworthy in the cause of rare breeds, the *MasterChef* competition on television was won by Silvija Davidson, who based her recipes on rare breeds. She was followed by other finalists in the competition using Southdown lamb (1992) and Tamworth pork (1993). A few years ago, breeders of *Black Welsh Mountain sheep* made a marketing ploy of their breeds: they not only sold excellent meat but also identified it closely with the breed name, which is a rare phenomenon in today's butcher's shops and supermarkets. Down with anonymity! And during the 1980s a *British Harvest restaurant* at the Hilton hotel in London featured named rare breeds on its menus, when Richard Wear (of Ruslin Ryelands) co-ordinated the supply of lamb carcases from different breeds.

Then there is the more obvious marketing niche for *wool from rare breeds*, many of which have either coloured fleeces or unusual fleece qualities such as the lustre of the longwools or the fineness of the Shetlands. There is now a project to market North Ronaldsay wool from the island, for example, and the Trust has injected starter finance and a loan to get the scheme going. Several individuals have specialised in rare-breeds wool, from Anne Kingham in Surrey, who used mainly Jacob wool in the 1970s, to Sheelagh Holmes (Island Heritage) who specialises mostly in Manx Loghtan and North Ronaldsay wool.

The promotion of rare breeds is the final piece in the jigsaw and increasingly marketing is being identified as a critical ingredient in the recipe for the success of the rare breeds. Enterprises such as Heal Farm Meats reaffirm the importance of emphasising special characteristics and high quality but there are difficulties inherent in the marketing of rare breed products. Precisely because the breeds are rare, production volume is low and lacks continuity. Such problems have been intensified by the closure of small local abattoirs and the continuing erosion of the market for small speciality butchers by large supermarket chains. These developments have jeopardised the marketing of rare breed products and the Trust, prompted by Richard Lutwyche, is taking a long look at the possibility of setting up a *marketing co-operative*. A successful marketing policy will provide the security and stability needed to encourage more farmers to breed rare breeds and the stimulus to define their characteristics and qualities more clearly.

But in the exhilaration of finding a commercial niche for rare breeds here and now, it should not be forgotten that the Trust exists to conserve rare-breed genes *regardless* of commercial value.

. Rare Breeds Task Conference, October 1971. Peter Jewell presents his paper. The meeting was chaired ▸y Sir Dudley Forwood Bt, flanked by Ann Wheatley-Hubbard (left) and Christopher Dadd (right)

. Chairman of the Trust, Michael Rosenberg, and fund-raising officer, Marshall Watson, discuss the Trust's ◂hibit at the Royal Show in 1982 with Minister of State for Agriculture, Peter Walker (later Lord Walker) ▸hoto - Peter Adams)

Above: 3. Denis Vernon presents the trophy for the champion Lincoln Longwool ram, Billingborough Caesar, to Harold Nobes BEM, shepherd for breed chairman, Robert Watts, at the Volvo-sponsored National Show and Sale in 1988
(photo - Simon Tupper)

Left: 4. Interbreed pig judge, Geoffrey Cloke, and Viki Mills with one of her Berkshire entries at the National Show and Sale in 1986. Viki took the Interbreed Championship with a Middle White gilt
(photo - Simon Tupper)

5. Michael Rosenberg receives the CBE from Minister of State for Agriculture, Michael Jopling, in 1986. Also in the picture, left to right, are Denis Vernon, Sir Richard Cooper Bt, Lord Benstead (Minister of State MAFF), Alastair Dymond, Joe Henson, John Hawtin, John Hearth (Chief Executive of RASE) and Geoffrey Cloke

6. HRH Prince Philip being welcomed on the Trust stand at the Royal Show in 1989 by Her Grace The Duchess of Devonshire. Staff members Val Nicholson and Kay Burgess join in the welcome (photo - Simon Tupper)

7. A reception at the Trust's demonstration unit near entrance 3 at the National Agricultural Centre. From right, J.D.F Green, Christopher Dadd (foreground), Shell U.K representative, John Lightbown (at rear), Michael Rosenberg, Mary Dadd, Richard Ferens (Hon. Director of RASE), Ann Wheatley-Hubbard (photo - Mike Norton)

8. Paul Hooper of Masterbreeders displays a tube of boar semen from the Trust's flask at a workshop in 1987. Geoffrey and Jonathan Cloke are on the left of the picture (photo - S.J.G Hall)

9. The owners of the winning pair of Longhorn cattle are presented with the Burke Trophy by HM The Queen at the Royal Show in 1981 (photo - *Farmers Weekly*)

10. Anne Petch, South-West regional winner of the NFU Mutual Marketing Award Scheme in 1984, receives her award from Minister of State for Agriculture, John MacGregor

11. A group of dignitaries at the 1991 AGM at Beamish. Left to right: front row, Denis Vernon (Treasurer), Mrs Teresa Wickham (Director of Corporate Affairs at Safeways), Lord Elliott of Morpeth (retiring President); back row, Sir Richard Cooper, Bt. (Chairman) and Sir Derek Barber (later Lord Barber of Tewkesbury) (President-elect) (photo - Pat Cassidy)

12. Anne Petch being greeted by the Prime Minister at 10 Downing Street in 1987

Above: 13. Pat Cassidy, Editor of *The Ark* since 1983, at her desk in the Trust's new offices at the National Agricultural Centre
(photo - Eileen Hayes)

14. Council member Anne Petch (left) and Technical Consultant Lawrence Alderson (behind) at the 1991 National Show and Sale with Silvija Davidson, winner of the 1990 Masterchef competition with a rare breeds recipe, and David Natt   (photo - Simon Tupper)

5. Members of Council at a meeting in January 1981 at 35 Belgrave Square line up in front of a historic painting of the RASE. Left to right: seated, Robin Otter, Lawrence Alderson (Technical Consultant), Michael Rosenberg (Chairman), Denis Vernon, Christopher Dadd; standing, John Hawtin, John Cator, Christopher Marler, Richard Cooper (later Sir Richard Cooper, Bt), Peter Jewell, David Steane and Denys Stubbs (photo - Ruislip Press Ltd)

6. Sale of rare breeds at Bite Farm in 1983. Nancy Briggs in the ring (photo - Simon Tupper)

17. The AGM in 1988 was held in the delightful setting of Temple Newsam Estate, one of the most important centres in the Trust's Approved Centre scheme. Breeds were paraded for the large audience in a custom-built amphitheatre (photo - Yorkshire Post Newspapers Ltd)

18. Crowds throng the Trust's marquee at the Great Yorkshire Show in 1982

Above: 19. HRH The Duke of Gloucester inspects the Trust's major demonstration in the Farming Heritage feature at the Devon County Show in 1975. He is hosted by Lawrence Alderson (left) and Michael Rosenberg (photo - Jerome Dessain & Co Ltd)

Right: 20. Michael Heseltine lends his voice to the Trust's fund-raising effort at the Celebration of Food & Farming event in 1989 in Hyde Park. Alongside are Sir Richard Cooper, Bt and Her Grace The Duchess of Devonshire

21. Chairman of the Trust, Dudley Reeves, greeting HRH Princess Alexandra at the Trust stand at the second Hyde Park event in 1992. Merchandising officer, Rosalind Ragg, is in the background (photo - Simon Tupper)

22. A rare opportunity to relax at the National Show & Sale; Show Director, John Hawtin, and Technical Consultant, Lawrence Alderson, in 1985 (photo - Simon Tupper)

23. Wyllie Turnbull judging a British Lop class at the National Show & Sale in 1979. Geoff Collings is parading his entry, Geoffrey Cloke is the ring steward, and John Backhouse is in the centre of the group of spectators (photo - *British Farmer & Stock-Breeder*)

24. Left to right: Sir Dudley Forwood, Bt, Ann Wheatley-Hubbard (Chairman of the Trust), John Hawtin (incoming Director) and Michael Rosenberg (retiring Director) at the National Show & Sale in 1982 (photo - Simon Tupper)

25. An Approved Centre inspection workshop held in 1992 at Norwood Farm, owned by member of Council, Cate Mack. Left to right: Geoffrey Cloke (vice-president), Brian Brooks (chairman of Farm Parks sub-committee), Dudley Reeves (chairman of the Trust), Lawrence Alderson (Executive Director), Peter King (Field Officer), Jeremy Roberts (member of Council) and (seated) Andrew Sheppy (member of Council) (photo - Robert Terry)

26. A family group at Wimpole Home Farm; managed by Bernard Hartshorn (centre) and Shirley Hartshorn (right), it maintains important units of many rare breeds including Longhorn cattle and Bagot goats (photo - *Cambridge Evening News*)

27. The Trust's Patron, HRH The Prince of Wales, opened the International Conference at Warwick University in 1989. Here he is welcomed by Denis Vernon, Chairman of the Trust. Also in the picture are John Hodges (left) and Sir Richard Cooper, Bt

CHAPTER SIX

COMMUNICATIONS

LARGE BLACK SOW

SOAY

CHAPTER 6:
COMMUNICATIONS

However good the product, it is worthless to others unless they know about it; that is to say, unless its value has been communicated. Good communication is essential to the work of the Trust — and not just to tell the outside world about rare breeds. Good communication begins *within* an organisation, between individuals and between groups of individuals.

In his pseudonymous column in *The Ark* introducing the journal's third volume in 1976, Noah reported on the Trust's first Annual General Meeting on 30 December 1975, which had been held at the British Museum (Natural History) in London. He remarked on the excellent attendance, with members travelling from as far away as Cornwall, North Wales and Yorkshire. He suggested that the degree of interest shown in the discussion period indicated a positive sign of vitality and went on to say: 'Further encouragement can be drawn from the fact that although there was considerable debate on several points, there was no sign of acrimony. We take this as a sign that the Trust can be "large" enough to accommodate varying opinions, without sinking into the morass of factionalism so prevalent in many Societies.'

A little later (January 1978) Noah returned to the subject, which was clearly beginning to worry him. Under the heading 'A Potential Conflict', he wrote:

> At the risk of interjecting a serious note into the holiday season, we are increasingly concerned about a divergence of views within the groups interested in various aspects of breed conservation. At the outset let us clearly state that we do not have a 'side' in the matter. The breeder is, quite naturally, concerned with the 'improvement' of his or her stock. The academic, for want of a better word, is concerned, quite properly, with the maintenance of a wide range of 'genetic variability'. In numerically small breeds there is no real conflict, because breeders are forced to use all available blood-lines regardless of the qualities those blood-lines represent.
>
> But, with those breeds which are more firmly established, problems begin to arise. Breeders are bound to 'select' for something, even if they do so in a negative manner. They are also bound to 'select' against certain undesirable characteristics and outright defects. Unless a critical blood-line is involved it would be very hard to convince a breeder to retain animals with characteristics he or she consider 'undesirable'. An animal of an historically correct 'type' may be of great interest to the scientist but hold little appeal to the breeder who is often subject to the pressures of the contemporary market place.
>
> The affairs of many of the breeds with which the Trust is concerned are operated by Breed Societies. These organizations are, quite properly, concerned with the 'promotion' of their breed, again an area of possible misunderstanding and conflict of interest.
>
> We feel that the only way to prevent a serious rift in these areas is to maintain a continuing dialogue between all parties involved. The pages of The Ark are always available for this purpose, and we hope that 1978 will see the expansion of other opportunities for discussion of this nature.

Sometimes, internal personality clashes have led to conflict within the Trust but that is to be expected in a group of people who, by the very fact of their apparently perverse delight in non-commercial breeds of livestock, are bound to be highly

103

individual. These are no followers of fashion; they are eccentrics in the sense that they do their own thing and it is noticeable that a high proportion of them answer to no boss: they tend to work for themselves. And they are passionate about rare breeds. Sometimes the Trust's communications network with its members has let it down, in spite of best intentions. This seems to be especially the case with a fairly vocal minority: the support groups. They perform a valuable function within the Trust and act as its agents in all parts of the country, but occasionally a group will strain against the central leash and feel that it is being 'ordered' rather than consulted.

The subject of this chapter is vital and extensive, and can broadly be divided into two sections: indirect communication on paper (journals, books, information sheets, articles, illustrations and so on) and on the air (radio and television); and then face-to-face in the form of talks, meetings, conferences and workshops. They are all part of the Trust's work.

ON PAPER

The Ark

The most important communications organ for the Trust has always been its highly respected monthly journal, *The Ark*. It was intended as such from its inception: it was a medium through which all the members were kept in touch with Trust activities, philosophies and ideas but it was a two-way process. Members could express their own views, in the letters pages or in articles, and could exchange livestock by means of the advertising columns. In addition, the magazine had always been a tool for recruiting new members to the Trust. For a full flavour of the Trust throughout its history since 1974, read all the volumes of *The Ark*.

To most people, *The Ark* is a well-loved old friend. Its contents seek to keep a balance between the highly technical information needed by researchers and breeders, the more basic husbandry guidance needed by the inexperienced, and general items appealing to a much wider readership — breed histories, personality profiles, descriptions of farms and centres and so on — as well as basic information about meetings and shows, and a bit of philosophy to serve as a debating point in the often energetic letters section.

Over the years, *The Ark* has developed into an essential archival source of reference for anyone interested in livestock breeds and has also become a journal of considerable quality in production and content. It has played a very important part in building up the Trust's good reputation as a respected organisation with a serious purpose and a sound approach to its task. In recent years it has become even more visually attractive since the introduction of colour for its cover.

The Ark was conceived by Michael Rosenberg and Lawrence Alderson as a private venture. Their aim was to provide a communication forum for the Trust and its members, a monthly journal, in conjunction with a show demonstration programme and promotional programme to assist the Trust by expanding its membership. It was, it must be emphasised, a private venture and for several years it remained as such: it was not produced through the Trust itself, but from farm sheds in the wilds of Devon with the help of volunteers.

In March 1974 Alderson had reported to the Trust's Advisory committee that an anonymous donor would provide up to £2,500 to launch a monthly magazine as a private exercise but devoted to rare breeds and obviously therefore of enormous help to the Trust. Far from being delighted at such generosity in publishing a smart little monthly which promoted the rare breeds movement, the committee's members objected. They did not like the idea of anonymity; they were unhappy that the magazine would not be under the Trust's direct control; they worried about tax liability, and the implications for the committee as custodians of charitable funds if the donor suddenly died. Finally John Cator stood up and said that he, personally, was prepared to stand surety in the sum of £2,500 against anything going wrong, and on that basis Alderson was given the authority to proceed in his negotiations with the donor.

By September, Alderson was able to report that *The Ark* was 'becoming established as a valuable line of communication in rare breeds survival work and many readers have expressed their appreciation of it'. It was being mailed regularly to 1,120 addresses, already a substantial increase from the 444 names supplied to Rosenberg by Hunt in the spring of 1974. Although it remained a private publication, it had been a valuable medium for recruiting members to the Trust and in advertising animals for breeding exchanges. Yet there were still twitterings of doubt among the committee: the publication's cover declared that it was the 'monthly journal of the Rare Breeds Survival Trust' and they preferred that the wording should be amended to include the words 'in support of', as they did not consider it to be the Trust's official organ. However, they decided to save themselves the expense of issuing the Trust's annual report and accounts by having them published in *The Ark*.

The first issue had been published in May 1974. On its cover was a photograph of a White Park bull named Charlie, from Henson's Cotswold Farm Park. An article written by Noah (a pseudonym that featured in *The Ark* for several years and hid the identity of its main instigator and financial supporter) introduced the magazine to its new readership:

> The past year has seen a great increase in the membership of The Rare Breeds Survival Trust. There is also an ever increasing interest on the part of the general public in the Trust's activities, and the plight of endangered species. There has been much media coverage of rare breeds, and rising attendance at Farm Parks and the Trust's exhibits at agricultural shows.
>
> With all this activity, we felt that the time was ripe for a monthly magazine about rare breeds. 'The Ark' is the result; we hope you will find it both informative and entertaining, and that you will enjoy watching its development.
>
> We do not want to start with apologies, but must say that the 'Ark' is not being compiled by professionals. It is the result of the combined efforts of a group of interested members of the Trust. It will be distributed free of charge to all members, and available for sale to the general public. Any profits from the sale of the magazine, or advertising therein, will go to the Trust, without any charges other than out-of-pocket expenses.

He then went on to describe what it was hoped would be continuing features: each month would see a different breed on the front cover with a related article inside which would combine background information on the breed with items relative to its progress. There would be articles about farms involved with rare breeds —farm parks, breeding centres and individuals who kept rare breeds as a hobby. There would be a monthly article in a technical vein; the first was by Alderson (on conservation versus breed improvement) and others would follow by authorities such as Barbara Noddle (a famous 'boiler of bones') and the sheep expert Dr Michael L. Ryder. The official voice of the Trust would be heard in a regular report from the Secretary. A selection of readers' letters would be published. The Trust's new registration programme for breeds which did not have their own society was deemed to be a high priority and would be covered regularly in the pages of the magazine. There would be standard features such as listings of shows, breed societies etc, and a classified section open to all: 'Anything is fair game for a small ad. but commencing with the July issue, we will not accept ads for the sale of animals which are not registered.' It was also hoped that readers would write with suggested topics for the magazine, and with their comments on specific questions when these were solicited. 'A movement of this nature,' wrote Noah, 'can only be successful given the active participation by a broad cross section of the membership.' And he ended: 'We hope that you will enjoy it as much as we have enjoyed creating it, and that your support will enable us to continue for more than the forty days and forty nights of our namesake.'

The first issue was slim by today's standards, running to about a dozen pages. It was published by Countrywide Livestock Ltd and printed by Optima Graphics of Clyst St Mary, Exeter. The featured 'farm of the month' was Cogden Farm, home of Mr and Mrs R.F. (Frank) Bailey on the Dorset coast and home to Jacob, Black Welsh Mountain and Soay sheep as well as commercial Welsh and Welsh halfbreds.

There was a 'List of Farm Parks', which included only seven: Alexandra (Leigh) Park at Portsmouth, Bedford Settled Estate at Woburn, Cotswold Farm Park at Guiting Power, Drusillas Animals at Alfriston, Farway Countryside Park at Colyton, Highland Wildlife Park in Inverness, and the NAC at Stoneleigh (though the latter might have raised an eyebrow at being called a farm park). The display advertisements were for Ash Farm, Iddesleigh, Devon ('A breeding centre for endangered species. Selected breeding stock available or bookable ...' which, as it happens, belonged to Michael Rosenberg), the Cotswold Farm Park rare breeds survival centre ('The most comprehensive collection of rare breeds of British Farm Animals in the country ...' which was Joe Henson's venture), and Livestock Improvement Services of Eastrip House, Colerne, Chippenham, Wilts (which just happened to be the address of one Lawrence Alderson).

Listed inside the front cover were the members of the Trust's 'large family' of animals and some of their names (given here exactly as they were printed then) might raise an eyebrow or two:

SHEEP: Black Welsh Mountain; Cotswold; Dartmoor White-Faced (horned); Jacob; Lincoln ; Lleyn; Manx Loghtan; Norfolk Horn; North Ronaldsay (Orkney); Portland; Rhiw; St Kilda; Shetland; Shropshire Down; Soay; Wensley Dale; White-Faced Woodland and Limestone; Wiltshire Horn.
CATTLE: Blue Albion; British White; Dexter; Gloucester; Irish Moiled; Longhorns; Shetland; White Park.
PIGS: Berkshire; Brit Lop; Gloucester Old Spot; Lincolnshire Curly Coat; Middle White; Oxford Sandy & Black; Tamworth; Ulster White.

The first issue was mailed to fewer than 500 people and there were several unexpected consequences, not the least of which was an influx of correspondence: during the month that followed its publication, letters were received from more than 600 individuals. This flood totally floored Rosenberg, who even today is not a good correspondent — and this was well before the era of word processors and other time-saving tools. Replies were often pecked out during a late-night watch in the lambing season, and the little post office at Winkleigh thought it had found a gold mine.

Most of the contents of the early issues were written by Noah and Alderson, and the journal was actually put together by Rosenberg and George Kayzar. Linda La Croix, one of the Trust's first life members, helped for nearly two years in the preparation of copy for the printers, and, like other backroom volunteers, she addressed envelopes, stuffed magazines into them and did anything she was asked in order to ensure the journal went out on time. Members and others began to contribute articles, letters and occasionally some fairly wild though enjoyable theories about genetics, breed origins and relationships. Gradually, over the years, the contents and the production became more professional but the magazine still managed to retain its friendly feel and to cater for a general as well as a specialist readership.

In 1977, the May Council meeting was informed that an anonymous American foundation had been set up with the purpose of making grants for rare breeds: a sum of £100,000 had been set aside for the Trust's projects over the next five years and would be invested in order to produce an income which would, specifically, secure the future of *The Ark* and also help with the costs of servicing members insofar as there was a shortfall from their subscriptions. In 1978 there was an editor's notice in the February issue:

> Due to the recent realignment of the upper eschelons of the 'Admiralty', Noah has found himself temporarily enmeshed in a position best kept free from partisan debate. During his absence members will be kept abreast of his less controversial doings through the columns of the Secretary and Alderson. The ARK would, however, be a 'dull ship' without a bit of controversy. The following is, therefore, the first of a series of signed 'Guest Editorials' which will take the place formerly occupied by Comment.

And, following Noah's deliberately provocative footsteps, the first of these editorials was by the equally provocative Devon vet, Jim Hindson, who enjoyed himself enormously at the expense of the 'traditional' and often obscure phraseology used in breed descriptions. Noah, though, seemed to sink without trace: his column was never seen again. But his influence continued: he was, of course, Michael Rosenberg, though his identification was never formally admitted.

In the autumn of 1979, while *The Ark* was still being funded entirely by individuals, its production office moved to Haltwhistle, as did the Trust's office.

The Ark has benefited from the long service of its few editors. The first was Noah himself; within a couple of years Alderson had taken over the editorial role while Noah continued to contribute his monthly column. Then in 1980 they handed the whole operation over to the Trust and the issue for September that year carried a new byline under the editorial: 'Management Committee: Lawrence Alderson, Judy Urquhart, Denis Vernon. Editor: Lawrence Alderson. Assistant Editor: Judy Urquhart. Editorial address: Market Place, Haltwhistle, Northumberland.'

Judy Urquhart had actually been brought in for the March issue (which included a brief fashion feature with photographs of Herdwick and Swaledale woollen garments) as Alderson's understudy and it was hoped that in due course she would become the journal's editor.

The September issue carried another allusion to the potential for creativity from conflict. Its editorial made two points in this connection. First of all it looked at the ever newsworthy conflict that seemed to exist between town and country, between working farmers on the one hand and conservationists on the other, and it saw the Trust as a useful meeting point between the two interests: its members had a responsibility to demonstrate the ability of different groups to live in harmony and to overcome ignorance. Then it continued:

> Much of the work in conservation is involved with the reconciliation of conflicting attitudes and requirements. In The Ark we must cater for both the experienced farmer and the inexperienced recruit from a different industry, for the scientific reader and the practical exponent, for the commercial breeder and the dedicated idealist. Which brings us back to the need for greater understanding and for education.
>
> This need is evident particularly when rare breeds seek a place in the livestock industry. They suffer from the strong resistance to change that afflicts much of farming industry ... their merit alone is unlikely to ensure their success. They will need a programme of publicity and education to

erode the barrier of preconceived ideas that bars their path.

And yet another item of interest in the 1980 volume — in August there was a large advertisement:

> A Secretary is required to establish and run an office to administer a charity with 3,000 members devoted to the conservation of minority breeds of British farm livestock. The location of the office is subject to discussion with the successful applicant. A salary of £7,000 is being offered initially for a job requiring regular office attendance and travelling on Trust business. Experience in administration, a knowledge of the livestock industry, an interest in conservation and the ability to get on well with people are all important facets of the position. Further details available from: The Chairman, Rare Breeds Survival Trust, Ash House, Iddesleigh, Winkleigh, Devon.

The Chairman was Michael Rosenberg, and in an open letter in the same issue he explained:

> During the past ten months the Council and Management Committee have conducted a searching review of our methods of administration. As a result of that review it has been decided to make significant organisational changes including the appointment of a Secretary, as advertised elsewhere in this issue. It is felt that these changes will enable us to more efficiently manage the Trust's affairs and to ensure that the work of Breed Conservation is encouraged and expanded in a constructive and orderly manner.

The Chairman already had it in mind that the Trust should home in on Stoneleigh as its headquarters. Fortunately the Shorthorn Society was willing to share its offices and *The Ark*, along with other Trust paperwork, moved there in 1981. An editorial committee was set up (Alderson, Rosenberg and Urquhart) and at the same time the more professional printing and production of the journal were taken over by Alec Paris Publicity, a public relations firm in nearby Banbury. The magazine was seen as part of the Trust's PR work. An Ark subcommittee, chaired by Richard Cooper and reporting to Council, was given the role of retaining continuity of policy and standards and also keeping an eye on the magazine's finances.

In April 1982 Mrs Pat Cassidy joined the team and took over the day-to-day office work in support of the news, editorials and features which were being solicited and collated, largely by Alec Paris Publicity. Cassidy, a farmer's wife for 23 years, had been keeping Portland sheep but otherwise had no obvious relevant experience for her initial job of working in the 'good ideas' department of the Trust — she simply knew Rosenberg through the Trust and was looking for a job. For six months she worked part-time but by 1983 she was effectively the editor, although her name did not appear formally as Editor until 1985.

In 1989 Alec Paris Publicity, who still produced the magazine, began to use desktop publishing techniques for layout, sending computer disks to type-setters and using Salvo Print for page paste-ups. That was the year when they introduced colour on the cover in celebration of the Festival of Food and Farming at Hyde Park. In 1990 production and printing of *The Ark* moved to Geerings of Ashford, Kent.

Since 1983 Cassidy has seen the magazine increase from 32 to 36 regular pages, and often 40. Production has improved noticeably. The editorial section now occupies nine pages; the greater number of features are much more varied and less ponderous. Circulation figures relate directly to membership and in 1993 the print-run was therefore 10,000 and increasing. *The Ark* is sent to all Trust members and goes all over the world. Many important libraries and places of education keep its volumes as reference material. There is no longer an Ark subcommittee but a much more functional editorial advisory group, monitored by and answerable to the Executive committee.

Livestock Heritage

The Ark has always tried to serve the Trust but it is difficult to please everybody. At one stage there was strong pressure for a glossy yearbook and in 1981 an ambitious magazine called *Livestock Heritage* was published. Its initial print-run was set at 15,000 — about five times that of the monthly and of the membership at the time. Intended as a fund-raiser, it was priced at £1 a copy and was sold from Trust stands at ten major shows in 1981 and also at farm parks and elsewhere. It had a full-colour cover with a Whitefaced Woodland ram painted by Gerry Acton and borrowed from Massey Ferguson's 1980 calendar.

The editor was Ann Spano in London and there were some good in-depth articles which appealed to a wide readership. There was a foreword by HRH The Duke of Edinburgh, commenting that at first sight the Trust 'would seem like another manifestation of the contemporary fashion in nostalgia' but that its real purpose was entirely practical. Article subjects included the North Ronaldsays on Linga Holm, the history of the Whitefaced Woodland 'sheep of the Peaks' (with a side article by the Duchess of Devonshire about the Chatsworth Woodlands — she was the Trust's vice president and past president), the past and present of the 'Old Chestnut' or Suffolk Punch (by Philip Ryder-Davies), the spinning of Wensleydale fleeces (by David Brewer), Celtic cattle (report of an Alderson paper presented to the third joint Scottish symposium), Middle White pigs (by Moreton Thomas, who had also written about the Whitefaced Woodland), rare poultry (by Fred Hams of the Rare Breeds (Poultry) Society), Longhorn sires, and a little trumpet-blowing 'Streets ahead!' about how the FAO had commended the Trust at an international event. There was also a directory section covering farm parks, breeders, breed societies and project sponsors.

Books

Quite apart from publishing *The Ark*, Countrywide Livestock had also published, or reprinted, a handful of books that would not have been considered commercial by others. They were of interest to Trust members, though: a history of Jacob sheep by A.J. Werner, and the Speed papers on the Exmoor pony came first. Then in 1983 the Trust (essentially Rosenberg and Alderson again) reprinted an Elwes book, originally published in 1913 when Elwes had set up a demonstration of primitive sheep breeds and their crosses at that year's Royal Show. The Trust revived the idea 70 years later, coincidentally with the opening of a permanent Rare Breeds Pavilion on the NAC showground, and printed the Elwes book again but with additions: there was an article by Rosenberg on the Combined Flock Book, and one by Peter Jewell as a recognised authority on primitive sheep, and there was a new final chapter by Alderson on 'The current status of primitive and rare breeds', which gave the up-to-date situation of all the breeds described by Elwes and added a little about one or two others.

The classic modern book about the rare breeds movement must be Alderson's own *The Chance to Survive,* first published in September, 1978 and reviewed by Rosenberg in *The Ark* in December that year. Other books have been written about rare and minority breeds; some are mentioned in the Bibliography, and reviews in back issues of *The Ark* give useful guidance on their quality. Then there are books which include rare among other breeds, perhaps the finest example being *200 Years of British Livestock* by two long-standing Trust members, Juliet Clutton-Brock and Stephen Hall. This superbly illustrated book includes many old paintings of rare breeds.

Other relevant publications include the proceedings of various conferences and symposia which promoted the Trust's work indirectly.

Leaflets and illustrations

Alderson also became involved in the production of wall charts illustrating farm livestock, ensuring that the rare breeds were duly given ample representation. The charts were published by Frederick Warne Publishers Ltd in 1979 and included six rare breeds of sheep, six of pigs and five of the 15 cattle. *Farmers Weekly* published its own wall chart of European livestock breeds, including some of the rare ones, illustrated by Jake Tebbit, the weekly's regular cartoonist but also an expert livestock artist who painted the world's cattle and pig breeds in Porter's books. The Trust published its own wall chart of the rare breeds in 1991, illustrating 43 breeds in colour.

The Trust also issued various information packs about its work and its rare breeds, aimed largely at non-members, and in 1983 Alec Paris conceived a concise little series of facts sheets, one for each accepted breed on the priority lists, which neatly encapsulated basic information about their history, performance, uses and appearance, accompanied by a black-and-white photograph. In 1990 the Trust produced a booklet, *Rare Breeds Facts and Figures*, which pulled together all the breed fact sheets.

The fact sheets had not been an entirely new idea. Back in 1976 the Trust announced its publication of the first three in a series of six-page leaflets giving the fully history and description of its breeds — initially Soay sheep, Tamworth pigs and Longhorn cattle. Unfortunately the Longhorn cattle society's chairman John Brigg wrote to the Trust with several complaints about the representation of his breed, including that an 'unsigned breed of the month feature' in the August issue of *The Ark* and also on the more recently published breed leaflet had been 'produced without the courtesy of any approach prior to publication to the society'. (At the same time he also disliked the fact that a pair of Longhorn bullocks had appeared at the Great Yorkshire show yoked to a cart, 'giving accent to the curiosity of the breed' rather than emphasising its valid commercial potential.) At that stage the Longhorn and one or two other breed societies were certainly rankled by what they described as the 'somewhat cavalier' attitude of the Trust and suggested that they should 'as of right' be represented on its Council.

On behalf of *The Ark*, Paris had accumulated a rapidly growing library of photographs. Some were old ones generously donated by members and breeders, others were new ones taken on farms and at shows by specialists such as Eileen Hayes, Sally Anne Thompson, Simon Tupper and, later, Sarah Rowland. This visual library is important and at intervals the Trust has appealed for more illustrations, especially where the subject is a good representative of the breed at its time and is properly identified. There are other photographic collections, above all perhaps the one at MERL (Museum of English Rural Life, now known as the Rural History Centre) which houses a huge collection of agricultural history on the campus of, Reading University. The museum is currently on a major fund-raising exercise to build more (and more modern) storage space for its thousands of glass plates, negatives, prints, books, old journals, herdbooks, documents and so on.

Several artists have specialised in illustrating the rare breeds: the animals have, after all, great visual appeal and such pictures tend to sell well to the general public. *The Ark*'s covers have sometimes featured paintings, including superb colour prints of 19th century breeds.

FILMS AND BROADCASTING

Rare breeds began featuring in television programmes right from the Trust's early days and in 1976 the Trust made its own television film entitled *Domestic Dodos*. This was on the initiative of Michael Rosenberg, who approached the BBC centre at

Plymouth and negotiated the making of a programme on the work of the Trust, within the *Access* series. Its producer was Brian Skilton and it was written and presented by Alderson during the summer of 1976, for showing in October that year. The twelve people featured in the programme included three Council members of the Trust, several members deeply involved in keeping rare breeds, and a few farmers whose main interest was in the commercial potential of some of the rare breeds. The half-hour programme was recorded on video-tape and this was purchased and converted into film with the help of a £500 grant from the Ernest Cook Trust so that copies could be produced and used in support of talks and demonstrations all over the country. In fact the Trust had launched an appeal to its members, partly for sponsorship for that specific film project and partly for individual donations to an Audio/Visual Aids Fund. Peter Hunt remembers running the film when he gave talks here and there about the rare breeds movement.

Over the years there have been various occasional televised items featuring rare breeds by Alderson, Porter and several breeders or farm park owners, especially that irrepressible showman Joe Henson. In the spring of 1975, he was approached by a leading BBC natural history producer for a documentary on the evolution of British farm animals and was asked to present a film which looked at the animals through the eyes of a natural history cameraman (the well known Maurice Tibbles). It was essentially a look at social behaviour patterns similar to those of their wild ancestors. It took a year to make this *Barnyard Safari*, which was shown in June 1976, and as well as the Cotswold Farm Park animals there were contributions from Andrew Sheppy as secretary of the Rare Breeds (Poultry) Society and Charles Martell as secretary of the Gloucester cattle society, both of them enthusiastically supporting the work on rare breeds survival.

In 1976 Henson had become involved with Johnny Morris's *Animal Magic* television series and other television projects, including scenes at the RBST Show & Sale in *Out of the Rut* and several involving his farm park, which ultimately led to a programme about rare breeds and some of the people who owned them — people like George Styles (Gloucester Old Spots), John and Maureen Latimer (sheep), Charles Cottrell Dormer (Longhorns at Rousham Park) and Andrew Sheppy (poultry). From then on Henson found himself quite frequently in front of the cameras, which he admits to enjoying thoroughly, and during 1976 he was on a world tour for a BBC film about domestic animals ranging from cattle and horses to camels and elephants.

Filming at the Trust's Show & Sale is popular with television companies. Arthur Anderson, producer for Scottish Television, made a programme in 1990 that was shown throughout Scotland and the Borders. Claire Powell, a good friend to the Trust, brought Welsh cameras to film in 1992 for the regional farming programme, 'Down to Earth', and in 1993 the Show & Sale featured on 'It's a Vet's Life'.

There has also been a short 16mm film, 'The Rare Breeds of Britain', made by the John Deere Corporation of America.

The breeds and the Trust have frequently been the subject of radio features. Again, 1976 seems to have been something of an early bumper year: for example, Denis Vernon was one of several who featured in broadcasts by Radio Carlisle (and on Border Television) during that year, making sure that the Trust was duly mentioned. There have been many other mentions and features over the years — on general programmes such as 'Women's Hour' as well as 'Farming Today', and often in the 'On Your Farm series'. Clearly the idea of rare breeds appeals to the serious farmer as well as a general public that simply likes the idea of unusual looking animals with, in some cases, rather odd habits. The success of the seaweed-eating North Ronaldsays' saunter down Fleet Street back in 1973 proved that.

EDUCATION

In the mid-1970s the Advisory Committee set up a working party to think about education. The members, apart from Alderson and the Secretary, were D.P. Billings (head of the department of agriculture at the Village College, Bassingbourne), science teacher R.I. Bullock, G.S. Boatfield (principal of the East Suffolk Agricultural Institute), J.D. Guiterman (regional education officer of the RSPCA), R.E. Smith (teacher for rural and environmental studies at Tunbridge Wells Comprehensive School) and J. Thomas (deputy head of Donneston School, Sedgely, Worcestershire). Unfortunately this project died and the idea was not revived until 1991, when Cate Mack secured the approval of Council to develop a programme to include rare breeds within the national curriculum.

WORKSHOPS

The new Trust had very soon realised the importance of offering breeders a series of venues where they could learn more about their breeds, in very practical terms, and also meet other breeders to exchange experiences and breeding stock. They were designed to take the place of the Technical Consultant's visits to individual breeders, as these were imposing a prohibitive demand on the limited time available.

The workshops were informal occasions and successful at an educational level and also on a social plane, an important factor in that livestock owners are scattered all over the country and often work in comparative isolation. They also generated a sense of involvement in what in the early days felt comfortably like a big family, its members sharing a common interest and a feeling of knowing everybody else in the movement. It was to the benefit of the breeds that such people should meet and mingle, and the workshops provided a more intimate setting than, say, the annual Show and Sale. The workshops were held in different parts of the country so that they were accessible to as many members as possible.

The formal programme of Trust workshops began in 1976, the two that year being held at Ash Farm and at Bite Farm. The Rosenbergs' *Ash Farm* at Winkleigh, Devon, had been the heart of so much — *The Ark*, the Combined Flock Book, the Show Demonstration programme. It was also an approved centre. Michael and Marianna Rosenberg had bought the farm in 1973 and quickly established a multibreed sheep unit (mainly Hebridean, Whitefaced Woodland, Shetland, Portland, Manx Loghtan and Norfolk Horn; also Leicester Longwool, Wiltshire Horn, Shropshire, Wensleydale, Ryeland, Boreray, North Ronaldsay and Castlemilk Moorit among the rare breeds, and Herdwick, Gotland and Merino as well). They also raised various rare breeds of cattle (Longhorn, Shetland, Kerry and Highland), Tamworth pigs, and Golden Guernsey, Bagot and Angora goats.

John and Maureen Latimer's *Ravendale Farm* at Higham Gobion was another popular venue. Latimer, one of the Combined Flock Book's registration and inspection committee, kept various sheep breeds and was a prominent breeder of Dexter cattle. The farm featured in Thames Television's *Today* programme in January 1977, with starring roles for the Jacob, Shetland, North Ronaldsay and Portland sheep and Maureen's Dexter housecow.

Ken and Nancy Briggs's *Bite Farm* at Trimpley, near Bewdley in Worcestershire, featured frequently on the workshops agenda. Collecting rare breeds of sheep became something of a passion for the Briggs family: they would spend their holidays exploring islands and remote regions where native breeds could be found. For example, in 1976 they went to the Isle of Man on the spur of the moment, with the intention of meeting every breeder of Manx Loghtan in their own informal survey of the breed, and came home with one farmer's entire flock (a ram and nine ewes). The following year they went to the Shetland Islands ... But before the Manx arrived their 140-acre holding already had breeding flocks of Soay, Shetland, North Ronaldsay, Whitefaced Woodland, Black Welsh Mountain, Lleyn and St Kilda (Hebridean) sheep, and a New Norfolk Horn ram for their commercial ewes and a grading-up flock of Portlands — and a flock of Jacobs, three Tanfaced Welsh Mountain horned ewes, some Badgerfaced Welsh Mountains and, on their high hill farm in Radnorshire, a small flock of very hardy Herdwicks. They were serious farmers: their sheep were a business as well as a hobby. They originally ran commercial flocks (Mules and Swaledales), and a few Jacobs for fun, until they happened to be holidaying in Devon in 1973 when the chance came to buy a handful of Soays. And after that they couldn't stop.

In 1974 Briggs visited the specialist sheep event at the NAC (which was transferred in later years to the Three Counties showground at Malvern) and purchased a pedigree Bluefaced Leicester ram from Alderson. The conversation between the two men turned to other breeds and Alderson mentioned the Norfolk Horn breeding-back programme and the Linga Holm project. Briggs was hooked! He purchased some Swaledale ewes for Alderson at Hawes for the Norfolk Horn programme, and he took a group of the North Ronaldsay sheep that were brought to the mainland that year. He has been a close and valuable member of the rare breeds movement ever since.

Appleby Castle, in Cumbria, was the headquarters of Ferguson International Holdings Ltd., of which the Trust's hon treasurer was chairman. It was the most impressive of all the workshop settings, up on top of the hill above the river Eden beside the big castle with its 11th century Norman keep, and it always seemed to provide fine weather as well. It was a very popular venue, offering a fine collection of not just farm livestock but also birds ranging from waterfowl to ravens, owls, cranes and flamingoes. In later years the manager Tessa Edwards co-operated with Ross and Avril Harrison and other members of the Dales Support Group to produce some excellent workshops at Appleby, and the last, in 1992, was another successful occasion: the afternoon was devoted to Eric Medway on the subject of card grading.

The first workshop to be held in Scotland was in 1977, at *Gallangad Farm,* near Loch Lomond, by invitation of Willie Macdonald, managing director of the well known sheepskin firm, Antartex. The farm had a flock of Gotland sheep, imported from Sweden in 1973, as well as Shetlands and a herd of White Park cattle. The workshop was the result of efforts by the Scottish group organised by Richard Allan.

The Workshops programme had been handed over to the regional support groups by the mid 1980s and perhaps it faltered a little. There was an important workshop or seminar set up by the Trust with ADAS and the Agricultural Training Board in 1987 on the subject of oestrus detection: the Trust was deeply concerned about the apparent wastage of rare breed semen from low conception rates which, it discovered, resulted from poor management rather than from poor semen quality.

The old style of a full, centrally planned programme of workshops was revived at the end of the decade when Alderson became Executive Director. There was, for example, an excellent one on Geoff Walter's Abbey Farm at Stixwould, where he and the Lincolnshire support group put on a superb display of rare breeds of sheep. Then there was a truly memorable workshop at the NAC, Stoneleigh, organised by the two Devonian queens of rare breed pigs, Anne Petch and Viki Mills, in conjunction with Christianne Glossop of Pig Genetics who talked about pig fertility and graphically demonstrated her talk with the help of Martin Snell (son of Philip) blatantly garmented to represent a sow's reproductive organs. In 1993 an ambitious four-day workshop was held at the University of Liverpool, ranging from DNA analysis to livestock handling.

The subjects and themes of the workshops vary. Some are devoted to individual breeds, and these are increasingly organised now by breeders' groups. Some have a particular theme such as wool colour or card grading, while some are more general genetics workshops.

FIELD DAYS AND ANNUAL GENERAL MEETINGS

The Trust's AGMs and open days began as 'field days', the first of which was on 5 October 1973. Sir Dudley Forwood gave an opening address, followed by an address by Deryk Frazer of the Nature Conservancy. There were short talks on the problems of running a farm park (by Michael Ann of Drusillas, Mrs J. Forbes of Farway Country Park, and Peter Maunder, director of the Leigh Park venture owned by Portsmouth city council). These were followed by a 'brains trust' session to stimulate a general discussion, under the chairmanship of Sir Emrys Jones, recently appointed principal of the Royal Agricultural College at Cirencester but previously director of ADAS. Among those attending was solicitor Robin Otter of the Gloucester cattle society, who became a Council member of the Trust, giving plenty of good legal advice (he helped with the registration of the Trust's symbol as a trademark and for copyright purposes, for example).

The second Field Day was held in October 1974 at Portsmouth's Leigh Park, and would be remembered for Jim Hindson's speech on the Three Black Spots syndrome. The Earl of Cranbrook, as president, chaired the morning session and the guests were welcomed by the city's Parks Department, corporate members of the Trust. Sir Dudley Forwood, as the Trust's honorary treasurer, gave a financial report. During the afternoon the Trust's chairman Joe Henson showed a film about the Linga Holm project and there followed a panel discussion chaired by Bill Longrigg of MAFF with Juliet Clutton-Brock of the Mammal Department of the Natural History Museum, Mrs A.J.C. Seymour (the Dexter cattle breeder), Andrew Sheppy of the Rare Breeds (Poultry) Society, and Alderson.

The main address was given by Jim Hindson. He began with a practical and detailed assessment of the special problems of rare breeds, and then went on to detail the main areas in which he felt the Trust's activities should be concentrated (he was a regular suggester of ideas through the columns of *The Ark* in those days, often controversially). The most popular part of his talk was the final one on the Three Black Spots syndrome: 'Is there a danger that you may be inventing breeds?' he challenged, simply to make the list of the Trust's breeds longer? 'We read of White Black Welsh Mountain Cattle and of Black White Welsh Mountain Sheep, of six-horned Jacobs and of four-horned Soays. Is it possible that we are witnessing what has been termed "the three black spots syndrome" at work? That is, an animal born with three black spots behind its left ear — there is a fanfare of trumpets and, lo! another breed exists.' How much of that was true mutation and how much a trace of previous outcrossing? He was worried about the suggestion of the Northern Dairy Shorthorn as a separate breed; he questioned whether the Blue Albion had ever bred true even in the old days; he suggested that the Dexter was simply the 'heterozygous phenotype of a lethal dominant gene and so can never breed true', and that the Red Poll's 'hybrid origins are known and a large injection of Danish Red blood has recently taken place, so is it intended to preserve only those types which have not involved any outcrossing?' He admitted to being deliberately provocative with these remarks, and no doubt a little harsh: he simply wanted the Trust to pause for a close examination of basic principles, 'standing by those that survive the examination, and modifying or rejecting those that do not.'

The Natural History Museum, working home of Juliet Clutton-Brock, was a frequent London venue for the Trust's annual meetings, including the very first of the Trust as a company, in December 1975, when its Council members were actually elected as such. The occasion was linked with the Trust's Christmas lectures and there were two speakers, with the common theme of St Kilda: Alderson spoke on the island sheep, while Professor R. John Berry of the Royal Free Hospital medical school described the genetics of the island's mice.

The following year, at the same venue, the AGM speakers were Juliet Clutton Brock, who presented a display of Garrard's 19th century livestock models; Rex Woods, on keeping rare breeds of poultry; and Clifford Owen, deputy curator of the Leicester County Museums. This was the last AGM attended by the Earl of Cranbrook as the Trust's founding president: he was elected as vice president instead. The 1977 AGM was once again held at the museum, which played host on several occasions altogether, most recently in December 1984.

The headquarters of the Zoological Society of London was the venue on two occasions (1978 and 1980). For the first, the speakers included Bowman, Jewell, Hindson and Rosenberg. In 1983 the father of the whole rare breeds movement, Lord Zuckerman himself, gave a talk at the Trust's AGM.

Temple Newsam, Leeds, is the largest of the Trust's approved centres — its flagship, in fact — and is another favourite venue for AGMs. Its Director of Leisure Services, John Tinker, was already a Council member when he played host to the AGM there in June 1988. The AGM returned there in 1993, in conjunction with the City's centenary celebrations.

In 1987 the AGM was held at Wellington Country Park, Stratfield Saye, by invitation of the Duke of Wellington as the Trust's president, and a major display of rare breeds and rare-breed products was toured by HRH The Prince of Wales, the Trust's Patron. It was a very windy day indeed and after His Highness had departed some of the livestock marquees dramatically blew inside out. Less alarming AGMs have been held at Edinburgh Zoo (1985), the Cotswold Farm Park (1986), Stoneleigh (1989), Wimpole (1990) and Beamish (1991). In 1992 the AGM was held at Cricket St Thomas, near Chard in Somerset. This venue was chosen because of dignuntlement on the part of members from the south west of England, and it gave people a chance to air their views in a direct manner.

Sometimes AGMs generate heated debate, perhaps on personal hobby-horses more than Trust policies and actions in general. And from time to time there have been outbursts, when people have raised problems and attacked the Trust quite violently. For example, the redoubtable Mrs Ann Baker (wife of the Australian professor Clive Manwell — as a team they are well known for their papers on domestication) rose to her feet at one AGM to protest about the way in which a project on polymorphisms was being carried out, and at another she expressed her strong disappointment that the Trust had not taken the Welsh Cob within its remit. At another, Alan Cheese, an employee at Shugborough Park in Staffordshire, objected that the Trust was not doing anything about Irish Moiled cattle, Middle White pigs or the Blue Albion; he followed up on Longrigg's remark that no bulls had been licensed since the war by writing an article in *Farmers Weekly* stating that he had in his possession a list on MAFF notepaper showing that no fewer than 46 Blue Albion bulls were licensed between 1952 and 1973.

CONFERENCES AND SEMINARS

CAS Symposium, 1982

In July 1982 the Trust participated in a symposium organised by the Centre for Agricultural Strategy, held at the University of Reading (the university of the CAS chairman, Professor Colin R.W. Spedding). The theme was animal gene conservation and the occasion was part of a feasibility study on embryo transfer and collection for which sponsorship had been gained from Shell. The main sessions were physiology and techniques; the preservation of species and breeds; and the logistics of preservation. Among the papers was one given by the Trust's Technical Consultant on 'World Needs in Farm Animal Preservation'. Some of the other speakers, all highly regarded n their fields, were those who had already contributed to the work of the Trust — for example, Chris Polge from Animal Breeding Cambridge Ltd; Dr W.P. (Twink) Allen peaking on the preservation of horse breeds; Kevin O'Connor of the Milk Marketing Board; and Geoff Mahon of British Livestock, who was an embryo transfer expert — he had already worked with Will Christie at Vallum Farm in Northumberland and had successfully transferred embryos from White Park cows.

Joint Scottish Symposia

A series of Scottish symposia began in December 1975 at Edinburgh Zoo. These were joint events between the Trust, the Fauna Preservation Society and the Royal Zoological Society of Scotland that would continue annually for at least a decade. The theme in 1975 was conservation and there were five speakers: Alderson presented a paper on the Trust's work, and Bill Carstairs spoke about the North Ronaldsay sheep. Another paper was given by Roger Wheater, director of Edinburgh Zoo, who had co-ordinated the symposium. The chairmen were Lord Balerno, of the Balerno Trust which had provided money for the MMB/RBST semen bank, and Dr John Berry. The second annual symposium, on the theme of Cattle, was under the chairmanship of Professor R.V. Short of the MRC Unit of Reproductive Biology.

Policy seminars

Over the years there has been a handful of what might be termed policy seminars. The first of these was a joint conference in December 1978 at the NAC, Stoneleigh, on 'Minority Breeds in Commercial Systems'. It was organised by the RASE, the Agricultural Development & Advisory Services (ADAS) and the Trust and proved to be a landmark for the Trust in winning an argument. The seminar was chaired by Richard Cooper and there were papers by, among others, Rosenberg, Alderson, Bob Large and John King, director of ABRO. An article, 'The hidden promise of forgotten breeds', appeared in *Farmers Weekly* as a result, largely outlining King's talk and pointing out that he had previously refused to speak publicly on the subject of minority breeds but now felt the time had come to look at their values and problems. It was a clear signal that the Trust's work was being taken seriously. King had previously been of the opinion that there really was little point in genetic conservation, yet now he was speaking about the importance of investigating, experimenting and documenting results from the breeds that should be conserved. He went on to say that more would be achieved by closer co-operation between breeders and State technical services than by the 'setting up of museum-type parks' which merely maintained herds of rare livestock, and that State aid should be geared towards this end. The Trust still receives no State aid and relies entirely on membership subscriptions, sponsorship and private donations to finance its work.

The success of this conference encouraged the Trust to negotiate for another joint venture as soon as possible. It took place on 14 February in 1980 as a confidential, in-house policy seminar at Stoneleigh, the results of which would be conveyed to the Trust's Council. It had been inspired by Michael Rosenberg and organised jointly by Dennis Willey and Mike Cornwell-Smith of ADAS, George Jackson of the RASE and Lawrence Alderson for the Trust. Jackson was the seminar's very effective chairman (he was eventually co-opted to the Trust's Council). Dr Malcolm Willis of Newcastle University opened the first of six main sessions. He was followed by Dr Kevin O'Connor of the Milk Marketing Board, John Newton of the Grassland Research Institute, Dr John Hinks of the Animal Breeding Research Organisation (ABRO), Richard Wear (breeder of pedigree Ryeland sheep) and Professor J.M. Cunningham of the Hill Farm Research Organisation in Scotland, who later became principal of the West of Scotland College of Agriculture. Broadly, it was a seminar about conservation policy, and members of the Trust such as Geoffrey Cloke, Peter Jewell, Jim Hindson,

Richard Cooper, Bill Longrigg and others mingled with invited guests such as Keith Cook from the MMB (who had been so instrumental in encouraging the collection of semen from rare breeds of cattle) and Ken Baker of the Meat and Livestock Commission (who was later co-opted to the Trust's Project Development committee).

On 19 October 1985 there was a policy meeting at Appleby Castle to discuss a paper drawn up by Alec Paris which looked particularly at membership policies and other items. It was a crucial time for the Trust: Rosenberg was gradually withdrawing from day-to-day Trust involvement, Cloke was taking on the role of Executive Chairman and Dymond was due to join the trust in January. Dymond was invited to attend the meeting and made a presentation of his plans for the Trust. They also considered the random growth of the new support groups: Dr Richard Allan was concerned that there was unsatisfactory communication between the groups and the Trust and that some form of organisation would be useful.

WENSLEYDALE

CHAPTER SEVEN
SHOWS AND CENTRES

CLEVELAND STALLION

NORTH RONALDSAY

CHAPTER 7:
SHOWS AND CENTRES

It is at county and agricultural shows that the greatest numbers of the general public are likely to notice — perhaps quite by chance — rare breeds of farm livestock and, having noticed, become interested in the work of the Trust. Two major developments in the very early days of the Trust were the Show Demonstration programme and the Show & Sale itself.

THE SHOW DEMONSTRATION PROGRAMME

In the Trust's first year, 1973, the organisation was actively promoting itself and its breeds at shows with the help of Council members Bill Longrigg (at the Surrey County show in May) and Lawrence Alderson (at the Royal Cornwall in June). Both shows play their part in the Trust's history.

The show secretary of the Surrey County's bank holiday event was Stephen Lance. At the Rare Breeds Survival Task conference in 1971 he had suggested making available 'menageries' of rare breeds to be displayed at small shows in order to help the new movement. Lance, who publishes an annual Showman's Directory covering just about every event in the country, became involved in the co-ordination of rare breeds displays for the Trust. He would eventually lead the team assembled by Rosenberg to organise an ambitious display at the Food and Farming festival in Hyde Park in 1989.

At the Royal Cornwall in 1973, his future partner in setting up those displays made his entrance into the Trust's history. Alderson was there with a few pens of rare-breed sheep which caught the eye of a passing American couple, Michael and Marianna Rosenberg. They bought a few sheep, and so began a lasting and fruitful relationship between the Trust and Rosenberg.

From the outset the Trust was fortunate in that many show secretaries, in addition to being sympathetic to its aims, realised than an exhibition of rare breeds could provide an interesting and educational attraction for farmers and the general public. The Show Demonstration programme was expanded in 1974 to include the Royal, the Great Yorkshire and the Royal Smithfield shows, and Rosenberg urged the Trust to take on a full dress programme at these events for the 1975 season. Agricultural societies provided the Trust with free sites and tentage, and often a contribution towards the cost of livestock transport. Without the help of these organisations, the programme would never have grown as it did and the Trust's costs for promotion would have been vastly increased.

Through Alderson, Rosenberg asked the Council to allow him to create a series of display signs, printed professionally to give the exhibits a better appearance, and the Show Demonstration programme began to take the form that later became familiar all over the country.

The programme received its first big boost at the Leigh Park open day, when Stephen Lance introduced Rosenberg to C.F.J. (John) Hocken, secretary of the Devon County Show. Hocken, who was planning a Farming Heritage feature for 1975, wanted Rosenberg to become involved in his 'children's farmyard' attraction but Rosenberg convinced him that a large-scale rare breeds exhibit would be of far greater interest. The result was one of the largest exhibits that the Trust ever staged: it included a 120ft x 80ft marquee and a separate set of buildings to hold further livestock which appeared in a twice-daily parade. This space required much more display material than had been ordered originally and the printers were still ferrying in duplicate signs on the second night of the show.

Crowds flocked to the exhibit and word of its success quickly spread around the show circuit so that the Trust was invited to a full rota of events around the country. The first of many visits by members of the Royal family was at the Devon County in 1975, when HRH the Duke of Gloucester and Lord Mountbatten of Burma were given tours of the display. In later years this list would grow to include HM the Queen, HM the Queen Mother, HRH the Prince of Wales (who became the Trust's Patron), the Duke of Edinburgh and the Princess Royal.

Soon Rosenberg was in attendance at virtually every exhibit. While the programme focused on major events for reasons of efficiency and economy, he would also take this travelling circus to several smaller events each year (often at the urging of local members) including Melplash, St Mellons, North Devon and many similar venues. Touring the smaller shows was taken over in due course by local support groups and the Trust's Trading Company but in the early days the main unit was often called in to help. On one memorable occasion Rosenberg brought the entire display set-up from the Great Yorkshire to Tatton Park to assist Barbara Platt and the Lancashire Support Group, on his way to his next appearance at the East of England Show.

The effect of these displays was dramatic in several ways. First and foremost, they proved to be an excellent source of recruitment of new members. Many joined while they were at the shows and many more as a result of leaflets obtained at a display, or mailings made from a guest book, later replaced with a series of enquiry forms. The exhibits also proved an excellent vehicle for breed promotion and for publicising the annual Show & Sale. Even a partial list of the breeding units established as a result of contacts made at the demonstrations would be far too long to include here but Wimpole Home Farm and the Boxmoor Trust are two notable examples. Shows also provided an excellent chance to make or maintain contact with breeders, many of whom were present as exhibitors. Rosenberg, who stayed in a caravan behind the stand, developed a 'we never close' policy and morning coffee became a tradition, as did spirited late night discussions.

In the early years the format of each display varied widely, usually depending on the facilities made available by the host society. Major demonstrations were staged for special events such as the Bicentenary of the Bath & West in 1977, the Centenary of the Stafford in 1978, a special display of Pigs of the World at the Surrey County in 1981, and a Sesquicentenary at Lincoln in 1983.

However, year after year the biggest stand was always at the Great Yorkshire. After asking Alderson (a Yorkshireman) to exhibit a small collection of cattle in the cattle sheds in 1974, which was repeated by Rosenberg the following year, the organisers asked for a more comprehensive display to include sheep and pigs in 1976. Rosenberg's request for a 120ft marquee was granted and he made his plans accordingly. When he arrived at the showground, direct from the Royal, he found that they had provided two interlinked 100ft marquees instead: the showground contractors could not set up the 120ft model underneath some overhead power lines. Fortunately he had arrived on the Friday before the show and spent the weekend borrowing additional livestock and penning, and totally destroying the transport budget (which, to the benefit of the Trust, he was underwriting personally). He persuaded John Gall of the Beamish Museum to fill up some of the space with a display of farming artefacts and old photographs, and this became part of the exhibit for many years.

The Great Yorkshire exhibit was a great success and became a major feature of the show from 1977 to 1985. As it was usually the penultimate show of the Trust's season and entailed far greater numbers of staff than usual, the space also proved an excellent venue for an end-of-season party and there are many fond memories of Don Clinch barbecuing steak sandwiches behind the stand late into the night, often to the tune of a Northumbrian piper who was part of the Beamish entourage.

Such a large demonstration programme required a lot of help and Rosenberg found this in quantity from many sources. The chief contributors were Don Clinch and his sons, Geoffrey and Mike, who joined their father in running Reddaway's Stock Transport when they came of age. Each of the boys worked for several years as Stand Manager, unloading and erecting penning and displays, tending the stock during the show, and packing up when it was all over. Together with their father, they would also help with these thankless chores even if they were only there as hauliers, and they would also help to staff the stand when things were busy. By 1982 the scope of the programme had grown to the extent that Rosenberg enlisted Town & Country Promotions to transport and erect the displays, and Wendy and Rory Finucane helped with the livestock side in 1984 and 1985.

Over the years Rosenberg developed a long list of members who would come and provide real assistance on the stand, sometimes for a day or sometimes for a whole show. Without their help, the Trust could never have achieved such an impact on the agricultural show scene.

By 1984, shortly before he was forced to retire from active participation, Rosenberg was looking for a way to make the programme more efficient and cost-effective. He felt that a small exhibit, perhaps limited to one example each of cattle, sheep and pigs, would be ideal as all the penning and displays could be carried in a small lorry with room for a stand manager's living quarters. An early version of this was tried out at the Showman's Show in Newark in 1984, and Alastair Dymond refined the idea which was finally brought to fruition in 1986.

Trust stands also provided an ideal selling point for merchandise. On a visit to the Great Yorkshire in 1980, Rosalind Ragg persuaded Rosenberg to propose to Council that the Trust should become involved in this activity. It proved such a success that for several years it threatened to swamp the recruitment and enquiry functions of the major stands but this was brought under control by providing a separate area for those purposes.

An offshoot of the Demonstration programme, and one with far-reaching implications, was the Trust's participation in a promotional tour of livestock markets which was mounted by Volvo UK in 1982. A public relations firm approached Rosenberg in 1981 about a proposal they were putting to Volvo in connection with the launch of the new 740 Estate, which they wanted to demonstrate to farmers throughout the country. The resulting promotion, sponsored by Volvo, enabled the Trust to acquire their first small display caravan and this appeared at a series of markets throughout the winter and spring with Trust promotional material and merchandise, accompanied by a small pen of livestock (usually sheep). Its banner read 'Volvo Cares for Conservation' and the unit was adjacent to stands displaying Volvo cars. It was operated for the Trust by Don Clinch, who travelled widely from the West Country to Scotland, greatly extending the Trust's exposure. At the end of the campaign the caravan was handed over to Rosalind Ragg for her fledgling merchandising effort and for several more years Volvo generously provided a vehicle to tow the unit around the country. The sponsorship then passed to Daihatsu, who continued it until 1993 when it passed to Butler Fuels.

Rosenberg made his final appearance on the show circuit in 1989 when, together with Stephen Lance, he produced 'Our Living Heritage' (in association with the Trust) as part of the nationwide year of Food and Farming. The concept of this display had been born in 1984, when Rosenberg first learned of the planned events: he had convened several meetings at Stoneleigh to discuss the Trust's participation. The Trust's resources would not permit participation on the scale he suggested.

Rosenberg and Lance designed a series of vignettes, or stage sets, each illustrating a different period in history from the Neolithic up to modern times. Animals were presented in each scene, often with live actors playing the part of cavemen, shepherds or Georgian artists. The entire exhibit was housed in a two-storey 'geodetic structure', quickly nicknamed The Dome, which was taken to four national shows (the Royal Ulster, Royal Highland, Royal Show and Royal Welsh). It had its debut at the Festival of Food and Farming in Hyde Park, London, where it was visited by Her Majesty the Queen. Attendance was tremendous and Trust literature and merchandise were always on hand. At Hyde Park, Rosenberg and Lance made the entire structure available to the Trust and Richard Cooper used the premises for a cocktail party and charity auction, which he conducted with the help of John Thornborrow and Michael Heseltine.

The Show Demonstration programme complemented show-ring entries of rare breeds and in those early days the Royal Smithfield was an increasingly important venue. One of those present at the Trust's Field Day in October 1973 was A.S.R. (Bunny) Austin, secretary of the Royal Smithfield Club and also of two sheep societies — those for the Dorset Down and for the Black Welsh Mountain. At the time the latter was deemed to be a rare breed but its prospects were greatly boosted when Joe Dudley's Black Welsh Mountain sheep won the Hill Sheep championship at the 1973 Smithfield. From then on the rare breeds became an increasingly familiar sight at Smithfield, largely due to Rosenberg's determination that there should be adequate classes for them. He virtually flooded the place with entries of Hebrideans, New Norfolk Horn, Black Welsh Mountain and so on. By 1975 there were classes for Whitefaced Woodland and St Kilda (Hebridean).

Rosenberg's enthusiasm for Smithfield was shared by Richard Cooper, whose later series of Smithfield Dinners introduced many useful contacts to the Trust.

Rosenberg was the great driving force behind the Show Demonstration programme. He put all his energies into it, all his creativity and a very great deal of sheer hard work, and he did so right from the start, from his 1973 meeting with Alderson at the Royal Cornwall. He enjoyed the challenge immensely. A naturally excellent host, he saw the value of creating a major social occasion for which he produced champagne and barbecued steak, laying on parties at the major venues that really gave the programme a 'whizz and a bang' in true Rosenberg style — and it was all highly productive as well as enjoyable. He used those parties to twist the arms of potential corporate members and donors, as well as to recruit individuals to the cause. During the period from 1980 to 1985, the membership of the Trust trebled; but in the absence of Rosenberg during the next five years it increased by only 30%. The combination of his magnetic personality and style (and not a little cash) made for success. And he enjoyed it all so much that when he began to reduce his Trust activities in the mid 1980s he asked to be allowed to persist with the Show Demonstration programme.

Today the Show Demonstration unit continues under the direction of Peter King and Mike and Jenny Clinch, visiting ten major shows each year, while Rosalind Ragg's mobile exhibition unit attends more than 40 smaller events.

THE TRADING COMPANY

From its early days the Trust had been selling little knick-knacks by mail order through *The Ark* or on the spot at shows. The earliest example was probably the promotional car stickers devised by Peter Hunt in 1973 for an open day. In 1976 the Trust first experimented with its own Christmas cards, essentially as a promotional exercise: they depicted White Park cattle clustered around a newborn calf and were entitled 'Animal Nativity'. In 1977 the Trust acquired a small holding in Joint Charity Card Associates Ltd, which marketed under the trade name Help Cards. In that venture the Trust found itself linked with major charities such as Dr Barnardo's, the NSPCC and the RNLI.

Then there was the Trust necktie — not to be belittled. Professor Colin R.W.

Spedding, head of the Agricultural and Horticultural departments at Reading University, wore the tie when he went into a lecture hall in Christchurch, New Zealand, in 1977. Somebody asked him about it; he told them a little about the Trust and thought no more of the incident. A few weeks later he was sent a cutting from the *Christchurch Times* which devoted nearly eight column inches to the Trust's work.

Those 'Old British Livestock' ties had first been advertised by Hunt in *The Ark* of February 1975. In those early years the ties, along with T-shirts, breed leaflets, breed origin charts and books, could be bought from Hunt or through the magazine's editorial office at Winkleigh, Devon, although throughout the seventies it was all a rather low-key and amateur business. Upjohns sponsored some Trust paperweights in 1979 but many of the other items were sale-or-return goods supplied by various members, particularly by Mrs Gill Henson from her Cotswold Farm Park shop.

In December 1980, with Rosalind Ragg's persuasion, the Council's authority was given to spend up to a thousand pounds on merchandise for selling on behalf of the Trust. At that stage everything tended to be stored at Ash Farm, the base for the Show Demonstration programme, or at Appleby Castle, the home of the Trust's treasurer. On 1 February 1981 the Trust opened its offices at NAC, Stoneleigh, and the whole focus shifted there. However, there was no room for merchandise or for all the back issues of *The Ark* which were heaped up in sheds at Ash Farm. Instead they went to Foxcote Farmhouse, the Warwickshire home of Rosalind Ragg, who had already been roped in for a breeders' workshop in March 1978 to talk about goats on behalf of the British Goat Society.

During that 1981 season she found herself camping out at shows all over the country (accompanied by her two small sons) and she managed to sell over £10,000 of merchandise that summer. She needed every volunteer she could find to help man the stand and it was a excellent chance for members to meet locally and begin to form their own regional support groups. From then on her life became a frantic (but highly organised) chaos.

In 1982, Volvo generously produced a vehicle and the Trust was able to purchase a mobile exhibition unit to go with it, so that she could travel even more widely — more shows, more livestock markets, more events here and there, all the time generating interest in the Trust and support for its work all over the country. The subsidiary RBST Trading Company Ltd was formed in April 1982 (in order to comply with tax regulations) and its three directors were Rosenberg, Vernon and Ragg, with the latter responsible for all buying and selling.

A more suitable towing vehicle was bought in 1988. Meanwhile the trading venture kept growing, until by early 1991 Foxcote was bursting at the seams with Trust merchandise. A portakabin was put up at the Trust's Stoneleigh offices and most of the merchandise was moved up there, along with all the trading company's office equipment, so that, for the first time, nearly everything was at last under one roof.

Geoffrey Cloke became a director of the Trading Company in 1984; Rosenberg retired in 1988, and Terry replaced Vernon in 1993. Rosalind Ragg had outlasted her original co-directors and is still there today, working as hard and as successfully as ever — a reassuringly familiar face at so many events.

THE SHOW & SALE

Noah, writing for his column in July 1974, explained that considerable space in the magazine had been devoted in the summer months to reporting the results of rare breeds in the show-ring — deliberately, as it was felt this important aspect of the movement had been relatively neglected. It had been discovered that a surprising number of classes for rare breeds existed but new ones needed to be established. To that end the Combined Flock Book registrations were being studied to see which breeds would be the most likely candidates for showing in 1975, over as wide a geographical range as possible. Because good venues were important, *The Ark* planned to 'offer a contribution to prize money when a site is selected'. The recent success of the Black Welsh Mountain at Smithfield had been a real boost and the 'possibility of similar results for other breeds, plus the advantages of public exposure, and the opportunity presented for breeders to get together and objectively compare their results, will more than justify the effort needed to establish classes of this nature.'

It might be fair to say that it was the success of the Leigh Park field day in October 1974, where 42 livestock breeds (many of them endangered) and 60 varieties of poultry were on display, that inspired the Trust to plan its very own Show & Sale for 1975: it was a logical development in the light of Noah's comments.

The Show & Sale committee set up to assist Alderson consisted of Christopher Dadd and Richard Cooper; they would be joined later by Andrew Sheppy, who would virtually carry the entire poultry programme on his own shoulders.

They had been planning it for months. Alderson had given details of the preliminary arrangements to the March committee meeting and by April John Thornborrow & Company had been appointed as official auctioneers on Dadd's advice: they would donate half their net commission to the Trust. It was anticipated that there would be 250 sheep and 50 cattle, and it was hoped that the Trust would at least break even financially. Its sources of income, apart from its share of the auctioneer's commission, would include entry fees and donations for prizes; it was felt at that stage that gate fees would be too difficult to administer.

Although it was necessarily a fairly small event on that first occasion, it attracted 1,500 visitors to Stoneleigh (in the end about half of them did pay to enter) and resulted in the sale of 183 out of the 321 sheep entered, and 34 out of the 52 cattle. The Trust had authorised the organisers to include poultry if they so wished but had decided against pigs, partly because of swine vesicular disease and partly because 'they require rather specialist treatment'. The RASE lent its support, providing penning at a reduced rate as long as the Trust could supply a team to set it up.

Ann Durham, Christopher Dadd's secretary at the time, still has vivid memories of watching the preparations from her office window. The animals were penned on the roadside by the collecting ring and she was amused by all the odd sheep in the pens, especially when Wally McCurdie, the NAC head shepherd, quite failed to contain the athletic little Soays or to dampen the excitable attempts of the Portlands to visit a Cotswold ram: every time the shepherd returned them to their rightful pens, they promptly vaulted out again. There was also the disadvantage of simultaneous dog shows and equestrian activities elsewhere on the site.

Alderson also has his memories of that great event. They had decided that they needed three sheep stewards, one cattle steward, a clerical assistant and six gate attendants who would function for an hour each, and it was hoped that as many Council members as possible would come along to fill those roles. In the event they needed 26 officials, three of whom were employed on a casual basis but the rest were volunteers. What a contrast in organisation to the Show & Sale in 1993, when the ten teams of helpers for the ten senior stewards totalled almost a hundred people.

The judges at that first Show & Sale included John Taylor (recently retired as secretary of the NCBA), Frank Bailey, Frank Houlton, C.A. Marton, P. Johnson (Chairman of the Longhorn cattle society), Rex Woods (past president of the Rare Breeds (Poultry) Society) and C. Bradbury. Interbreed judges over the years are shown in Table 7.1. The stewards were Ann Wheatley-Hubbard, Richard Cooper, John Cator and Andrew Sheppy. Michael Rosenberg provided the entirety of his Show Demonstration display material. Wally McCurdie stewarded the sheep classes; Nigel and Nell Maydew tackled the paperwork; Peter and Moira Hunt dealt with a steady flow of enquiries in the office; and Alastair Dymond dealt with the logistics of the sheep pens.

Table 7.1: Show & Sale Interbreed Judges

YEAR	CATTLE	SHEEP	PIGS
1975	J.A. Taylor	C.A. Marton	-
1976	G. Thompstone	R.P. Wear	J.W.F. Causton
1977	C.J.S. Marler	P. Furness	D. McLean
1978	J. Cumber	J.R.C. Rough	A. Sherriff
1979	R. Vigus	P. Mummery	T.H. Copas
1980	W. Longrigg	A. Veitch	A. Gregory
1981	Mrs M. Johnston	A.W. Lang	J. Howlett
1982	T. Best	G.L.H. Alderson	Mrs E.R. Wheatley-Hubbard
1983	C.B. Playle	G. Hughes	P.G. Barrett
1984	J. Rossiter	J. McNaughton	H. Rowe
1985	C.J. Hutchings	F.G. Ball	G. Walters
1986	R.E. Needham	P.A. Whitcombe	G.E. Cloke
1987	J. Weyman-Jones	J.R. Taylor	R. Overend
1988	G.M. Rankin	J.M. Johnston	Mrs C. Parker
1989	W. Young	J. Reed	R. Russell
1990	F.J. Williams	R. Boodle	R.T. Swanton
1991	Mrs D. Flack	R.P. Wear	-
1992	B. Hartshorn	J. Thorley	-
1993	C.J.S. Marler	J.K. Briggs	Mrs B.A. Petch

Modestly flushed with success after this first event, Alderson reported in *The Ark* that it had all 'provided an excellent medium for breeders to evaluate their own animals alongside those from other herds and flocks' — an essential exercise which enabled them to 'modify their policies as necessary and to identify the most desirable sources of breeding stock', which was its main purpose alongside generating enthusiasm in the general public. The final entry had been 45 cattle, 358 sheep, four goats and 198 poultry.

He was fulsome in his gratitude, particularly to the exhibitors themselves. There had been seventy of them, too many to list individually but he singled out Joe Henson and John Neave of the Cotswold Farm Park, Michael and Marianna Rosenberg of Ash Farm, and Ken and Nancy Briggs of Bite Farm, all of whom had 'supported the event with especially large entries'. He thanked Juliet Todd, Bryan Jones, Roger Brown and others who had helped the auctioneers on the eve of the show with penning and general preparations and, above all, the auctioneer himself: John Thornborrow, a stocky Cumbrian who specialised in pedigree livestock sales. Thornborrow operated his business for many years from offices in Leamington Spa but, after a brief association with Midland Marts, he returned to his native haunts and settled in Penrith in 1993.

The day, albeit hectic, had its humorous moments, 'although on occasion it was difficult to see the funny side at the time'. Alderson's carefully planned advertising programme, for example, had been 'methodically mutilated': in one of the farming journals the British White was listed with British Friesian, the Black Welsh Mountain sheep was listed with Welsh Black cattle, the Soay 'assumed a synthetic flavour under the title of Soya' and the St Kilda was classed as a 'multihorned pig'.

But the event had been a success in terms of publicity and as a display of rare breeds and a meeting point for Trust members and livestock breeders. Financially it was less satisfactory.

The first two events were organised mainly by Alderson and Thornborrow, and then for several years by Rosenberg, who began to hand over the baton to John Hawtin in 1981 for the 1982 event. Hawtin happened to be in the right place one day when Rosenberg was looking for an extra pair of willing hands, and he was promptly recruited as a steward, thus beginning his rise to chairman of the Show subcommittee, which was established with the organiser as its chairman.

In 1992 Hawtin was unwell and his chief steward Frank Bailey effectively ran an event that had by then become immensely more complex than the modest affair of 1975. It now needed whole teams of stewards for all sorts of different jobs — arrivals, departures, species and so forth — and teams of contractors. The Trust has greatly appreciated the help provided over the years by Boy Scouts, pupils from Braunton School, members of HM Prison Service and many others, including teams from the RASE itself.

The RASE had always worked hard for the Rare Breeds Show & Sale, and its co-operation and lowering of charges were vital to the early events in particular. It had bent over backwards to make sure they were a success. John Hearth, its Chief Executive, had written to Christopher Dadd in October 1977, saying that 'the last Sale seems to have been particularly successful and I am most anxious that the Trust continues to hold its sales here,' enjoining Dadd to tell the Trust how much the Society had 'enjoyed having them and how much we look forward to having them next year.' Dadd did just that and managed to reduce some of the Society's charges, writing: 'As you know, we are quite prepared to cut the hire costs to the bone and allow you to do almost all the labouring work yourselves if you wish in order to save double-time charges for our own men at the weekend.'

A worksheet produced by the RASE's Trevor Mitchell for the 1980 Show & Sale shows the huge amount of work that was needed even at that stage. The build-up began on Wednesday; the event itself (it had become a two-day one in 1979) was on Friday and Saturday; and the 'breakdown' took up Sunday and Monday. ('Will all those involved,' came a heartfelt plea from Mitchell, 'please remember that the Dairy Event build-up starts on Tuesday ...!!') Rosenberg would be camped on site in the Rank Village even earlier — from the weekend before the event. RASE staff would prepare and hand over on that build-up Wednesday a whole range of facilities: 'Cattle sheds 1—12; 6th Street; Foreman of Stock Hut; Area between sheds 3 and 4; Isolation boxes; Rank Hostel (115 rooms); Toilet blocks 4 and 5; Radio telephones; Barriers; Chairs; Tables; Headboards; Hurdles; Pig pens; Tiered seating; Poultry cages.' The cattle sheds would have been washed and disinfected; the hurdles, pig pens and poultry cages cleaned; and all premises cleaned. A great list of equipment had to be delivered and, where necessary, erected by the previous Tuesday evening, including four sets of tiered seats, 500 chairs, 110 tables, 75 barriers, 100 pig pens, 400 hurdles (over and above those supplied by the auctioneers) — and on and on went the list. They sorted out electricity supplies to all the sheds (including checking 'that ALL bulbs and tubes are working — Mr Rosenberg to obtain keys for electric plugs of his choice...') and hot-water facilities as well as cold-water standpipes. They put barriers across the roads to enable the Trust to collect gate money; they cleaned the toilets; they prepared the 'lorry wash' and organised the car parks (and attendants); they made ready for the arrival of livestock vehicles and prepared isolation boxes; they made sure that the RAC had put up signposts; they even provided 'four wide brooms'. Time and time again in this detailed worksheet the name of Alastair Dymond was mentioned: he at that stage was farm manager at the NAC but was described as 'the local representative on the organising committee, and any matters requiring a decision from the Organisers point of view should be referred to him.'

Over the years the number of trade stands at the show has increased. Space was first allowed for them at the 1976 event: Alderson had recommended their inclusion as a means to raise income as the first event had not been a financial success. By 1993 they would occupy more than three cattle buildings and the NSA building.

The poultry section has remained a little apart from the large livestock. It developed in conjunction with the British Waterfowl Association section in the Tate & Lyle building (replaced in 1992 by the New Exhibition building) until it was moved to the Cattle buildings in 1993. The BWA became involved because Alderson was its secretary/treasurer at one stage and Christopher Marler was also a focal figure. The birds add further interest and variety to an event that has become not only the major attraction in the Trust's calendar but also one of the largest sales of pedigree livestock in Europe.

Among the large livestock, the Cattle Interbreed Champion at the first Show & Sale was a British White bull, Hevingham Ferdinand. Since then the cattle classes have been dominated most recently by the Longhorns but overall by the White Parks, which have the most interbreed championships to their credit. The honours have been spread more evenly among the pig breeds. In the sheep classes the early dominance of the Down and longwool breeds, with a tradition of showing, has been usurped lately by champions from the ranks of the Hebridean and Portland breeds.

Some 'side' aspects of the Show & Sale have become features in their own right — for example, the hospitality unit. It was originally intended only for Council members but is now more democratically available to 'guests' willing to pay for their meal. Sir Dudley Forwood plays the host, standing at the head of the stairs to greet personally all those who climb them. Meanwhile down on the floor at the heart of it all would be Geoffrey Cloke drumming up custom for the Raffle. It is a role he relishes and he has become adept at loudly cajoling old friends as they try to sneak past with their wallets carefully hidden.

The Show & Sale continues to grow. In 1993 it was at its biggest ever, extending for the first time to minority as well as rare breeds, accepting both accredited and non-accredited health scheme animals, and spreading to occupy the sheep, pig and NSA buildings in addition to the cattle buildings. There were ten show-rings and six sale-rings for large livestock, and numerous judges and card-graders appointed by breed societies and breeders' groups. The honorary medical officers, Harper-Smith and Harrison, and honorary veterinary officers, Hindson and Watson, ensured that emergencies were dealt with promptly; Gavin Wilcock presided over the event from the public address system; and Don Clinch was a tower of strength around the site, as he had been in previous years in tandem with Ken Turner (who unfortunately was absent in 1993). It was all a baptism of fire for field officer Peter King, who had been with the Trust for little more than a year; he and Kay Burgess, as permanent members of the staff, co-ordinated the event but they relied heavily on the voluntary senior stewards in each department: Jonathan Cloke (pigs), who also acted as understudy and assistant to John Hawtin; Eric Medway (sheep); Jenny Taylor (fleeces); Peter Guest (cattle); Ian Kay (poultry); Roger Stokes (arrivals, and health and safety); Angela Lightwood (departures); Phil Roughton (traffic); and Robert Terry (trade stands). The senior stewards, in turn, were dependent on a large and dedicated volunteer force of stewards who worked long hours to ensure the event's success.

Naturally there have been controversies over the years: it is inevitable in a competitive situation. Perhaps the most heated and misunderstood has been over the system of card grading.

Card grading

Most breeders like the idea of showing their livestock. But there can be dangers in showing, especially where genetic resources are limited.

In 1980 Alderson circulated some notes on the showing of primitive breeds of sheep and exposed the problem of a lack of detailed type-standards for those breeds with no breed societies of their own. Breed standards, he posited, are used not so much by breeders selecting their breeding stock but are for the guidance of judges (and breeders) in the show ring. 'In most popular breeds,' he said, 'uniformity within the breed is a matter of high priority, and type standards are an important factor in achieving this uniformity.'

Showing also demanded that animals should be well presented, carrying more condition than would be normal in a breeding animal. The two factors of type uniformity and overfat condition in presentation both tended to lead to a reduction in genetic variability, a very undesirable trend in the conservation of primitive sheep breeds. He recommended that the latter should always be presented in their natural state in breeding condition and should be judged with maximum emphasis on functional characteristics and a sound constitution.

A couple of years later Alderson produced another paper, 'Showing of rare breeds and its relevance to genetic conservation'. He noted that the Show & Sale had been immensely successful but that there were certain inherent dangers. 'In particular the increasing interest in rare breeds and the influence of the show ring could lead to the creation of fashion points within a breed and to a hierarchical breed structure. The result of this would be to produce a greater degree of uniformity within a breed with the resultant loss of genetic variability.'

The question was, how could this potential creation of fashionable types be avoided? A combined defence was needed. First of all, judges for the Show & Sale should be selected from differing backgrounds; their briefing should emphasise 'those basic characteristics which are typical of each breed and which are associated with constitutional soundness'. Secondly, a rigid type specification for a breed should be avoided: there should be no leaning towards the show-ring disease of elevating minor and unimportant points based on breed standards. Thirdly, in the interests of preserving maximum genetic variability (the watchword of the Trust), breeders needed to be educated in the problems of inbreeding; the maintenance of all existing bloodlines could prevent domination by a single herd or flock in a breed.

At that stage, Alderson felt that 'these safeguards will ensure the genetic variability within each breed is maintained as far as is reasonably possible'. However, within two years there were some worrying trends. *The Ark*, in the first months of 1984, published a series of 'Getting prepared' articles devoted to the art of the show-ring. The authors were all highly respected in their fields: the delightful and very experienced Donald McLean on showing pigs, for example; the Ryeland breeder and shower Richard Wear on sheep; and on beef cattle Tony Carr, promotions director for the Museum of Cider in Hereford and previously herd manager of the prize-winning Crickley herd of pedigree Poll Herefords.

In response, Alderson published an article, 'To show or not to show' — and that was, indeed, the question. Shows, he said, were fun but were they of benefit to the breeds of livestock that paraded in the rings? Was there a danger that they might create a hierarchical breed structure and narrow the genetic base of the breed?

In November of that year there was a lively meeting on the subject of breed standards among a group of people from a wide range of backgrounds and attitudes. Alderson set the scene, explaining the current position and the problems that were apparent or imminent. Then he handed the floor to two people with differing opinions. One was Frank Bailey, who had been an inspector for the Combined Flock Book for several years; he bred Jacob sheep and enjoyed showing, and he proposed the uniformity approach — that the Trust should have strong and strict standards for its breeds. In contrast Juliet Clutton-Brock argued for as wide a variety as possible within each breed in the interests of genetic variation. Then the debate was thrown open and there was a very lively discussion indeed.

The general opinion seemed to be that there should indeed be *some* breed standards but that they should not be tight ones: they should simply debar animals that were not genuine examples of a breed. Then Geoffrey Cloke put forward a constructive proposal. He talked about a system already applied to poultry: instead of simply awarding rosettes to the 'best' entrants, judges would grade them into three broad groups based on their quality. All those in the top *group* would be awarded red cards; those in the middle group would be given blue cards and those in the bottom, yellow cards. The net result was that a group rather than one or two individuals became the focus of attention, and that allowed a much wider base for genetic variation.

The idea was accepted and was subsequently agreed in committee. This card-grading system was first tested at the 1986 Show & Sale and will become mandatory for sheep breeds in 1994. There are no plans at present to make it mandatory for other species but the idea was tried out for cattle breeds at the 1992 event and for pigs in 1993.

The problem has been to 'sell' the idea to breeders. In the early stages, Alderson and Frank Bailey would demonstrate the system, usually with Portland sheep, in various regions. There was, for example, an interesting session in Dorset on the farm of Norman and Michelle Jones: Michelle had bought a Portland ram and wanted to know exactly what its real quality was. Bravely she accepted the card-grading judgement that it was not very good!

The arguments continue, quite heatedly, especially concerning breeds which are not controlled through the Combined Flock Book (where the system was originally applied). Some have accepted it and used it effectively, such as Eric Medway and David Braithwaite as chairman and secretary respectively of the Hebridean sheep breeders group. Others are still resistant, perhaps through misunderstandings about how the system works. For example, there were misconceptions that the grading would be carried out by outsiders who did not know the breed, but in fact the animals are graded from within the breed's group or society. On the other hand, there are now those who would like to see showing of rare breeds abolished altogether, with the Trust having simply a card-grading and sale event; there are others who want 'show-only' classes as well as card-grading, and so on.

REGIONAL SALES

The tricky question of standards, set centrally by the Trust, arose frequently as local support groups developed. In the same way that the groups had grown up somewhat haphazardly, so too did local or regional sales of rare breeds begin to proliferate as a law unto themselves. When the Trust did attempt to exercise some control over standards, it often failed. There was something of an independent streak anyway, and even if standards could be set it was difficult to ensure that they were maintained.

The earliest local sales were in the mid 1980s — in York, for example, and Gloucester, Chelford and Kings Lynn. In some cases the Trust's recommendations have been accepted and a network of 'approved' sales is beginning to build up. Others have not been so co-operative; sometimes it is local auctioneers who resent the Trust's intrusion, sometimes a local support group insists on including unregistered stock, or even breeds which are not on the Trust's priority lists. The situation varies.

A set of rules was drawn up in 1992 covering the breeds that can justifiably be included in what purports to be a sale of rare breeds, the authenticity of the animals' pedigrees and registration, their conformance with the breed type, and their general health.

Why are rules necessary at all? Consider first the fundamental purpose of the regional sales, which is — or should be — to provide owners of rare breeds with an opportunity to exchange their stock locally, without having to travel long distances with individual animals that they wish to exchange. At a large event they can probably complete all their exchanges on one day, from a good selection of stock. It follows that such stock must be of good quality (and therefore need to be inspected) and properly accredited, i.e. a registration scheme is necessary. Above all it should be done in the interests of the *breeds*. That is the fundamental point of the whole rare breeds movement.

APPROVED CENTRES

The Trust's approval or otherwise of centres which breed or exhibit rare breeds is another area in which there have been a few disagreements along the way. In August 1973, shortly after the birth of the Trust, Alderson (as a member of the Advisory Committee) circulated a report which would lead to the Approved Centres scheme. He proposed two types: those involved in actual breeding work would be termed breeding centres, while others would be termed survival centres.

Long before the Trust was even an idea, all sorts of people (from the owners of great estates to smallholders) had kept collections of their favourite minority breeds. Some kept them because of a family tradition; some because they valued a breed's unfashionable merits; some wanted an unusual or handsome group as part of the landscape; and some kept them for no better reason than that the breed appealed, emotionally.

There were already a few places in the late 1960s and early 70s where collections of various breeds were on public display, and this highlights an early dilemma for the Trust. On the one hand, it was important to draw the breeds to the attention of the general public in order to gain more widespread support, and this could be done by inviting the public to farms and zoos where the breeds were kept, and also by ensuring that they were visible at agricultural and other shows around the country. On the other hand, it was vital that the low numbers among the rare breeds were boosted to ensure survival, and that needed to be done carefully and professionally as well as urgently. It seem that there were two distinct and not necessarily compatible roles for the rare breeds centres: public display, and pedigree breeding. Many felt that it was important for the Trust to have at least some control in both cases, in the former to ensure that the animals on display were true representatives of the breed and were in excellent condition, and that the facilities for the animals and for the public were the best possible, and in the latter (where public display was not the main aim) to ensure that a breed was not further endangered by slapdash breeding standards or management and could benefit from outcrossing with other lines on other holdings.

It was at the centres that the public had its closest contact with the Trust and it was therefore important that visitors should carry away a good impression of the centre and its animals, reflecting on the Trust itself, should the Trust agree to lend its name by approving a centre (in return for which it would have at least some control over the management of the animals).

In theory such approval was a good idea from the Trust's point of view but in practice it could present problems. For example, at some centres there were other activities which had no direct connection with the breeds, yet the latter were an integral part of the centre. It was felt, in such cases, that the Trust should look at enterprise as a whole, not just the animals, but should also look at the latter in detail.

In the early days, there had developed a feeling that farm parks were becoming too influential within the Trust and that commerce was overriding the scientific approach — the genetics approach. One or two fell by the wayside, or failed to meet the Trust's standards. In September, 1974, the Trust called a special meeting to consider standards in farm parks and survival centres and, after a lengthy discussion, the consensus seemed to be that some of the farm parks were below standard in their general management and in their displays of rare breeds. As they were acting as one of the Trust's shop windows for the breeds, it was important that their standards should be high in order not to discredit the work of the Trust.

Richard Cooper pointed out the distinction between farm parks and rare breeds survival centres. Farm parks did not have to have any direct link with the Trust, indeed did not have to be involved in rare breeds at all unless they wished; but the survival centres were specifically for rare breeds and thus must be asked to meet the Trust's standards. In other words, rules should be drawn up for survival centres but not necessarily for farm parks.

It was recognised that there was a delicate balance between the need to establish a 'group' policy for the centres but on the other hand to allow freedom of action for individual entrepreneurs both in commercial terms and in choice of breeding programmes. However, the future of the breeds was of paramount importance and the breeding centres must be prepared at least to record their livestock carefully and, preferably, to co-operate with Trust breeding policies designed to prevent inbreeding and to encourage the propagation of the best lines in the breed overall.

The centres were in need of guidelines anyway. Alderson was always happy to give advice to any keepers of rare breeds, and more specifically to advise prospective centres before they even began to think of seeking approval. If the Trust could establish a set of rules for the centres, then everybody would know exactly where they stood. Alderson was therefore instructed to draft what was required, and it was also agreed that he should be supported by an Inspection Panel to keep an eye on the approval scheme. Its original members were Michael Ann, John Cator, John Cole-Morgan, Richard Cooper, Sir Dudley Forwood, Deryk Frazer and Andrew Sheppy. It was recognised as important that the panel should include at least one member who was aware of the problems of coping with the public (for example, a farm park owner) and another who was a specialist in livestock management and breeding. The panel must be (and be seen to be) impartial in its judgements, and the constitution of the panel should remain as constant as possible in order to ensure continuity of the maintenance of standards. After approval, each centre would be inspected at least once a year.

As instructed, Alderson drew up a set of criteria for the inspection and acceptance of 'approved' rare breeds centres. At that stage the centres fell naturally into two types of ownership: those owned by organisations (such as the NAC and local councils) and those in private hands. Another division was between those that were open to the public and those that were not.

The criteria, which were approved at the meeting in January 1975, were based on three aspects: the management of the animals (including breeding policies and registration), the health of the animals, and the facilities for the public. These remain the basic criteria today. The principles remain the same but now a detailed questionnaire has been added.

Considering the livestock, the Trust's objective was to encourage centres to establish and maintain genetically pure, healthy breeding animals in groups of sufficient size to form an effective breeding unit and to conserve the characteristics typical of the breed. Health status was of prime importance, both for the survival and improvement of the breed and for the sake of a good display.

It was felt that the rare breeds survival centres would fulfil one or both of two roles: they could be conservation centres which maintained breeding units, and/or promotion centres which were open to the public. It was recognised that there might be different criteria for the different roles where the two were not combined.

To encourage centres to seek Trust approval, the carrots might include publicity through the Trust, the right to display an approved-centre plaque, eligibility for subsidies if specific Trust projects were undertaken, the ability to obtain and dispose of stock through the Trust, and priority for various services provided by the Trust.

Over the years those early guidelines have for the most part persisted and been developed. The Trust made full use of Christopher Marler's experience with his own public collection of everything from flamingos and peacocks to Ankole cattle and Big-horn sheep, but it never became as autocratic as the Zoo Federation. It also took advice from Brian Brooks of Hollanden Rare Breeds, who was deeply involved with the 'pick-your-own' industry and knew a great deal about the facilities needed by the general public.

In 1986 Alastair Dymond suggested that centres which were not open to the public should be dropped from the approved centres scheme and it was agreed to approve only those centres which were open to the public, dropping the breeding centres. The requirements for approval of centres open to the public would be based on the existing guidelines with the addition of a need to keep animals 'of a good breed type, free from defects, and a good advertisement for their breed'. It was important that the centre should create an overall favourable impression on the visitor and great care should be taken over the dissemination of information about the breeds. There should also be an area where the Trust's display boards could be on view.

By 1986 or so, owners of farm parks were meeting to discuss mutual interests on a regular basis. The group became an Approved Farm Parks subcommittee of the Trust's Council. Owners and managers of approved farm parks could send one representative to discuss common interests. The requirements laid upon approved farm parks were (and remain) detailed and quite stringent.

The Trust was aware that many farm parks had failed owing to a lack of proper planning and research, and by neglecting to prepare realistic financial forecasts. It therefore established, in 1991, a consultancy service to ensure that owners of proposed farm parks fully understood the implications and potential problems of their intended enterprise.

In 1992, the Executive Director of the Trust re-introduced the approval of breeding centres which were not necessarily open to the public but which were specifically places that performed a conservation role. They were designated as accredited breeding centres or units. The Executive Director was, of course, the man who had originally introduced the concept of approved centres back in 1973: Lawrence Alderson.

*

Some of the early centres failed for various reasons; others jogged along quite happily over the years; and others became of major importance to the rare breeds movement. Over the years there have been 41 centres approved by the Trust. They have been spread through all parts of the UK and in general there have been no more than one or two in any county, except in Devon, where there have been six centres. This tends to make the concentration in the south west a little heavy: there are fewer centres in the north of England and Scotland, and currently none at all in Wales.

The following thumbnail sketches give the flavour of most of those 41 centres. Four former ones have not been included as their policies proved incompatible with the Trust's guidelines and standards — they were Ashdown Forest Farm in Sussex, Babylun Sheep in Cambridgeshire, Cobthorn Farm near Bristol, and Farway Countryside Park near Honiton, Devon.

Aldenham Country Park
This is one of the younger approved centres and it lies within the M25 belt in Hertfordshire. The farm park, in which Longhorn cattle figure prominently, is only a small part of a much larger complex which includes activities such as fishing, sailing and an adventure playground. It is owned by Hertfordshire county council and managed by John Deacon.

Appleby Castle Conservation Centre
Appleby is one of the oldest centres and it surrounds the castle, the headquarters of

Ferguson International Holdings plc. It has a beautiful setting above the River Eden and has been a popular location for Trust workshops and think-tank sessions. Tessa Edwards manages the farm park and there is also a comprehensive collection of waterfowl, in addition to several important breeding units of rare breeds.

Ash Farm
Ash Farm was one of the early approved centres. Michael and Marianna Rosenberg built up an important collection of rare breeds — twelve breeds of sheep, three breeds of cattle and one breed of pig — in addition to breeding units of Highland cattle, Herdwick sheep and Angora goats. The centre was used to develop and prove new methods of livestock husbandry and progressive health programmes. The meticulous recording and grading system used for sheep has provided an invaluable reference point for the standards and performance of rare breeds of sheep. Ash Farm was also the location of some of the most successful of the Trust's breeders' workshops.

Bite Farm
Ken and Nancy Briggs also held some very successful breeders' workshops at Bite Farm, near Bewdley, Worcestershire. They kept Dexter cattle and a variety of sheep breeds but their main interest was goats. They played a major part in the Trust's early involvement with Bagot goats and they held important positions in the Golden Guernsey Goat Society.

Cluanie-an-Teanassie
This was the northernmost centre to be approved by the Trust and, like Ash Farm and Bite Farm, it was a breeding centre which was not open to the public. It was owned by Archie and Martha Crawford, who made the drastic transition from Shootlands Farm in Surrey to their new farm near Inverness. It later became well known for its herd of Red Deer.

Cotswold Farm Park
This centre is part of a much larger farming enterprise and is set in the middle of the Cotswold tourist area near Stow-on-the-Wold. It was established by Joe Henson and John Neave in order to tap the passing tourist trade more than twenty years ago. It has developed a range of rare breeds of several species, with a particular interest in local Gloucestershire breeds, and new facilities have been added over the years. It has featured frequently on television and radio programmes.

Croxteth Home Farm
Liverpool City Council established this centre in the middle of residential areas only five miles from the city centre, which can cause problems but also provides one of the largest 'gates', attracting more than 300,000 visitors a year. It is the former home of Lord Sefton and includes other attractions such as a walled garden, a miniature railway and an old cock-fighting pit. The centre has taken a special interest in Irish Moiled and Shetland cattle.

Dedham Rare Breeds Farm
Although this is one of the newer centres, Peter and Libby Harris have already built up an attractive display on the edge of the historical village of Dedham. They have a particular interest in rare pig and poultry breeds and are co-operating with the Trust to develop nucleus units for these two species.

Drusillas
Lying on the Sussex coast near Alfriston, this was one of the earliest centres. It was owned by Michael Ann, who developed it as much as a zoo as a farm park. It did not remain in the approved centre scheme after the early years but contact was re-established in 1991.

East Hele Farm
Owned by Richard and Anne Petch, this centre near Kingsnympton in Devon has become one of the most important for rare breeds of pig. In the course of developing her business of marketing the meat of rare breeds, Anne Petch has carried out much valuable investigative and trial work and the information she has amassed on pigs is comparable to that on sheep at Ash Farm.

Farmworld
Rare breeds form only a relatively small part of this farm park owned by the Co-operative Wholesale Society near Oadby in Leicestershire. There is also a major farming business. Perhaps the main attraction is the viewing of the dairy herd at milking time.

Gillhouse Herd
This is the second major pig centre in Devon and is owned by Viki Mills who, with Anne Petch, has played a major role in helping to give the conservation of pigs a high profile within the Trust's programme.

Graves Park
Owned by the City of Sheffield Recreation Department, Graves Park covers a large area over which the public roam freely. The centre left the approved scheme in 1992 but a group of the Trust's Bagot goats have been held there and the centre still maintains a herd of Vaynol cattle on the Trust's behalf.

Hatton Country World
This centre, managed by David Blower for Jonathan Arkwright, lies only a few miles from the Trust's offices. The farm park is part of a much larger complex which includes a large craft centre and a collection of vintage farm machinery. It was one of the first centres to hold an accredited unit of poultry (Croad Langshan) for the Trust.

Hele Farm
Owned by Hedley Le Bas, this was an approved centre in the early days of the Trust. It lay on the eastern slopes of Dartmoor and kept a range of breeds under the prefix 'Bolehyde'.

Hollanden Park Farm
Like several other farm parks, this is part of a much larger complex. It lies a few miles beyond the M25 near Sevenoaks, Kent, and is owned by Brian Brooks, who was chairman of the Farm Parks subcommittee for several years. The main business is a large pick-your-own enterprise.

Horton Park Farm
This centre, near Epsom in Surrey, is owned by Jackie Flaherty, sister of Sir Richard Cooper, Bt, and is managed by Caroline Key. They have created the park as a working farmyard with a wide range of rare breeds but with a special interest in coloured breeds of sheep. There is also a large playground which is very popular with local families.

Isle of Wight Rare Breeds and Waterfowl Park
Hugh Noyes has created a farm park in an Area of Outstanding Natural Beauty with splendid views. It relies heavily on the tourist trade, in the absence of any nearby major conurbation.

Leigh (Alexandra) Park
Owned by Portsmouth City Council and managed by Peter Maunder, this was one of the earliest approved centres. It dropped out of the scheme but not before it had hosted an important annual general meeting for the Trust in 1974.

Linga Holm
The Trust's refuge for North Ronaldsay sheep in the Orkney Islands was given approved centre status as an important breeding unit.

Longhouse
A breeding centre near Dereham in Norfolk, Longhouse is owned by Dr Peter Wade-Martins who has a deep interest in Manx Loghtan sheep.

National Agricultural Centre
The transfer of rare breeds from Whipsnade to Stoneleigh made this the first approved centre. Although not officially classed as a farm park, nevertheless it acted as a valuable shop window for rare breeds which were seen by many visitors to the Showground. The rare breeds of sheep have now been moved to other centres but the herd of White Park cattle remains and is part of the continuing link between the Trust and the Royal Agricultural Society of England.

Norwood Farm
This centre near Bath is owned by Council member Cate Mack, who has developed it in conjunction with an organic farming system. It has been the location for workshops in recent years and for a meeting of Approved Centre inspectors. The farm shop sells rare breeds meat and attracts customers beyond the visitors to the farm park.

Oban Farm Park
Kay Simpson, a qualified veterinary surgeon, opened this centre after she returned from Australia. It lies in a beautiful area on the west coast of Scotland and relies heavily on the tourist trade as the surrounding area is not heavily populated.

Parke Rare Breeds Farm
Tim Ash originally opened the West Wales farm park, where he became involved with the Ancient Cattle of Wales. He then moved to Devon and opened the Parke centre near Bovey Tracey under contract to the National Trust. It is on the main Exeter to Plymouth route and thus attracts tourists as well as local residents.

Ravendale Farm
John and Maureen Latimer's centre was yet another important site for breeders' workshops; in the late 1970s it became one of the 'big three' along with Bite Farm and Ash Farm. It maintained a large collection of the more primitive rare breeds of sheep and a herd of Dexter cattle.

Ridgway House
This early approved centre near Farnham in Surrey was owned by Michael and Anne Kingham. It developed from a flock of Jacob sheep and the garments created from their wool.

Riber Castle Fauna Reserve and Wildlife Park
Eddie Hallam, who owned this centre beside the ruins of the old castle near Matlock, Derbyshire, was a frequent visitor to Romania to study lynx. His interest in wildlife was evident at the centre but he also had a large collection of rare breeds and had acquired some North Ronaldsay sheep even before the Trust initiated its Linga Holm project.

Sandwell Park Farm
Situated close to Birmingham and owned by the Metropolitan Borough of Sandwell, this centre is similar to the living historical model farms seen in North America. It concentrates on those breeds that are native to the area and has also restored a large and impressive pigeoncote.

Sherwood Forest Farm Park

This Nottinghamshire farm park, situated on the edge of the Forest near Mansfield, is one of the more recently approved centres. The Shaw-Browne family has developed a wide range of attractions apart from the rare breeds, including water gardens and an aviary.

Shugborough Park Farm

Like Sandwell Park Farm, this centre has based itself on the living historical farm model. It emphasises old crafts such as butter-making, cheese-making and spinning and there is a working water-driven cornmill. It is owned by Staffordshire county council and is next to Shugborough Hall, the home of Lord Lichfield.

Temple Newsam Home Farm

Extending to more than a thousand acres, this is one of the most extensive of all the approved centres. It is part of the historic Temple Newsam estate, just east of Leeds, and is under the overall responsibility of Trust Council member John Tinker, as Director of Leisure Services for Leeds City Council. Farm manager David Bradley is a master-handler of bulls and significant breeding units of all the endangered breeds of cattle are maintained. It has been the location for not only a bull-handling workshop but also two of the Trust's annual general meetings, the second (in 1993) coinciding with the 100th anniversary of the Charter of the City of Leeds.

Tilgate Park Nature Centre

Owned by Crawley Borough Council in Sussex, this is one of the smallest centres in size (there are only a few paddocks) but not in impact. It is managed by Niall Osborne and is run in tandem with a wildlife centre. The policy of entering livestock at shows has paid dividends: their White Park bull, Dynevor Rampant, won 14 championships and reserves in a two-year campaign, beating bulls of some 30 British and Continental breeds in the process.

Toddington Manor Rare Breeds Centre

Although not open to the public on a regular basis, this centre close to the M1 motorway in Bedfordshire caters more for specialist groups and parties. Owned by Sir Neville and Lady Bowman-Shaw and managed by John Errington, it has developed important breeding units for several breeds, especially of cattle. It is also the home of one of the first accredited units of poultry (Croad Langshan and White Langshan) recognised by the Trust.

Wantsley Farm

Owned by Adrian and Elizabeth Seymour, this approved centre near Beaminster, Dorset, is important for its herd of Dexter cattle.

Wimpole Home Farm

This is the 'flagship' of the National Trust's involvement in rare breeds and genetic conservation. Close to Wimpole House, near Royston in Cambridgeshire, it is managed by Bernard Hartshorn, an experienced cattleman and respected judge who is currently chairman of the Trust's Farm Parks subcommittee. His wife Shirley is closely involved as well and their animals enjoy regular successes at shows. The centre includes a restored tithe barn and a museum of agricultural machinery as well as important breeding units of rare breeds, especially cattle.

Wye Valley Farm Park

Lying in a loop of the River Wye below the bluffs of Symonds Yat in Gloucestershire, this unit is centred round a farmyard and is part of a large farming business owned by Richard and Sue Vaughan. They have a particular interest in breeds of horses and poultry and also the genuine traditional breeds such as Hereford cattle. It was Sue Vaughan who first alerted the Trust to the perilous state of the pure strains of British Hereford.

APPROVED FARM PARKS

RARE BREEDS SURVIVAL TRUST

CHAPTER EIGHT
THE WIDER STAGE

LONGHORN CATTLE

BAGOT GOAT

CHAPTER 8:
THE WIDER STAGE

The conservation of traditional local breeds in other countries has been patchy. Some have become deeply involved, some entirely negligent; some have government-backed organisations, some are self-funding charities like the RBST. Some, indeed, have been inspired by the success of the Trust to form their own organisations, for example in the Netherlands, the USA, Spain and Switzerland. In some cases governments are actively involved in preserving what they see as a national heritage; in Italy, Spain and Portugal, for example, the Ministries of Agriculture publish beautifully produced books giving full details of even the most rare of their native breeds.

The Trust has had international links since its earliest days but in the present decade those links have become far, far stronger. Over the years its international status has increased to the extent that many other countries now look to the UK for advice in this sphere, and that in turn has enhanced the Trust's standing in its own country: it has long since ceased to be regarded as a vaguely tiresome and motley group of dilettantes and eccentrics, and is now respected and consulted by British agricultural organisations and by the government.

The international story really begins back in the 1960s with a couple of FAO study groups (1966 and 1968) on the subject of genetic resources for domestic animals. There was a fundamental difference in policy with that of the Trust: the FAO studies declared that breeds should be 'evaluated for their performance'. The Trust, believing that a breed's genetic material is irreplaceable (once the breed has gone, its unique genetic material is lost), acts on the principle that the breeds should be conserved for their own sake, regardless of their qualities, as insurance against the unknown needs of the future, rather than simply to meet current economic requirements.

In 1969 the increasing loss of genetic resources caused by more effective animal breeding programmes was discussed for the first time at a meeting of the European Association for Animal Production (EAAP) in Helsinki, when reference was made to earlier European work including the European Poultry Conference held in Bologna in 1964. The EAAP continued to take an interest in breed conservation thereafter; in 1980 its Genetics Commission set up a working party on the subject, which undertook a substantial survey throughout Europe (22 countries responded to its questionnaires). In its final report, published in *Livestock Production Science*, 11 (1984), this working party outlined the actions taken to conserve rare breeds in different countries; it identified 241 endangered breeds in Europe (which it listed, with populations); it recommended the creation of a central data bank on endangered breeds and it suggested a 'permanent organisation of AGR conservation activities for EAAP member countries'.

In 1972 there was an international UNESCO conference, 'Man and the Biosphere', held in Stockholm. It included 'the conservation of natural areas and of the genetic material they contain', and the theme was repeated 20 years later in Brazil at the UNCED Rio sessions. The Trust had not yet been officially formed at the time of that first biosphere conference: it was no more than a project being discussed by the Working Party, but this evidence of a growing interest in conservation on a global scale was an encouraging background for those discussions.

Even at the Working Party stage, there was overseas interest in the idea of the Rare Breeds Survival Trust. B.R. Gotto, of France's recently formed Société d'Ethnozootechnique, had heard about it through the Museum of English Rural Life at Reading University and wrote in April 1972 to learn more. The Trust's chairman and Alderson attended the Société's meeting in Paris early in 1975, and the French returned the compliment later in the year.

In 1974 John Bowman presented a paper to the first World Congress on Genetics Applied to Livestock Production, in Madrid. In April that year, the Trust was represented at an international archaeozoological conference in Holland and Alderson was in touch with Dr Anneka Clason, a close friend of Juliet Clutton-Brock's, who was a leading light in the rare breeds movement in the Netherlands. He helped her to set up Stichting Zeldzame Huiderrassen, and his fare to attend its inaugural meeting on 12 April 1975 was paid by the Trust.

In June the same year, Alderson embarked on a 20-day private business and lecture tour to North America, in the course of which he attended the ALHFAM (Association of Living Historical Farms and Museums) conference at Old Sturbridge Village, Massachusetts. He encouraged the Americans to think about their own rare breeds and advised a group which would found the American Minor Breeds Conservancy two years later. He also presented a paper at the university of British Columbia, in Vancouver, where (as it happens) one of the university's professors was John Hodges, who would eventually join the FAO and, on retirement from it, would become involved in the formation of Rare Breeds International as a trustee. The ALHFAM connection continued: Alderson gave a lecture to the Canadian branch in Ottawa in 1983, and returned in 1990 to give a paper at an AgCanada international conference. More recently Jy Chiperzak has established a separate rare breeds conservancy in Canada, where AgCanada (a government organisation) also hopes to play a leading role.

In 1979 Alderson was in Adelaide, South Australia, for the first Coloured Sheep Congress. This event was held thereafter every five years at different venues: New Zealand, Oregon and, in 1994, York (organised by the Trust).

1980 was the year of a major FAO global conference on genetic conservation, held in Rome, and both Alderson and Bowman gave papers. Alderson had visited Italy a year or two earlier when he was invited to give papers at an event in Reggio Emilio (Bologna) and to tour Tuscany looking at local cattle breeds. A decade later the Trust won the British Ford Conservation award and, represented by Vernon and Alderson in Cologne, was just pipped at the post for the European award.

Thus the Trust's international contacts began to build up, and other members have also fostered those contacts over the years — Richard Cooper, for example, as well as John Bowman. Geoffrey Cloke, who was chairman of the NPBA (and then the BPA) for nine years, became president of the European Pig Federation. Dr Rex Walters, a member of the Trust's Council and also the Trust's first representative on the Advisory Council of Rare Breeds International, has travelled widely as an expert in pigs genetics and is well known overseas.

More recently, a committee was called together in Paris in 1991 to make a survey of the continent's endangered ruminant breeds and to suggest conservation programmes. The Trust was closely involved in this, and has also undertaken a major survey of pig breeds in Europe on contract from the EC. Alderson has sat on a working panel in Hannover which administers the FAO/EAAP Global Animal Genetic Data Bank, supported by the EC Commission.

These international links are increasingly important, especially as EC legislation becomes more and more of a challenge to those who keep rare breeds — for example on semen banks and on slaughterhouses. Clearly the Trust will become much more involved politically in Brussels, and also at Westminster; already regular representations are made to Whitehall.

THE WARWICK CONFERENCE AND RARE BREEDS
INTERNATIONAL

In 1987, Alderson received a letter from Anneka Clason of the Netherlands. Almost as an aside, she suggested that perhaps it was time to create a European organisation of rare breeds groups. He took the idea further than Europe to include the Americas and other parts of the world.

The first step was an international conference, organised by the Trust, at Warwick University in 1989. It was well attended and there was an opening address by the Trust's patron, HRH the Prince of Wales. The proceedings were published later by CAB, and the conference produced one or two perhaps unexpected changes of heart within the Trust: it decided, after two decades of opposition to the ideas, to include within its remit certain traditional British breeds of utility poultry and it also agreed to take a little more interest in ferals.

Alderson used the Warwick conference as a launching pad for an international organisation. During the course of it he talked personally to many of the delegates, sounding them out on the idea, and then put a proposition to the final session. It was carried overwhelmingly and an acting committee was immediately set up under his chairmanship to create the new organisation. The committee members were Laurent Avon (France), Imre Bodó (Hungary), Roy Crawford (Canada), Hans Peter Grunenfelder (Switzerland) and Armando Primo (Brazil).

Alderson, as chairman, had the job of drafting the constitution, with the help of Fiona Middleton and Andrew Philips of the London solicitors, Messrs Bates, Wells and Braithwaite, recommended because of their experience with the World Wildlife Fund. The document was broadly based on the Trust's own constitution and the draft was accepted by the subsequent trustees at the 1990 meeting of the World Congress on Genetics Applied to Livestock Production. It was decided to change its name from the cumbersome title of the World Fund for the Conservation of Domestic Animal Genetic Resources; it was finally registered with Companies House in England as a charity named Rare Breeds International on 3 July 1991. The four trustees were Alderson, Crawford, Bodó, and John Hodges (who had just retired as the FAO's chief livestock officer).

In 1991, the Trust was represented by its chairman, Richard Cooper, and the highly respected Ian Gill of the University of Liverpool at the international conference in Budapest, during which the RBI held its first meeting and Alderson was elected as its founder chairman for a maximum term of two years. The 1992 conference was in Cordoba, organised by Cordoba University in conjunction with Spain's rare breeds organisation, SERGA, and the RBI to coincide with celebrations for the quincentenary of Columbus's discovery of the Americas. At that meeting, Hugh Blair (New Zealand) replaced Avon as a director.

In 1993, at Aarhus in Denmark, Alderson's term as chairman came to an end and he was succeeded by Crawford.

PAST AND FUTURE

The Trust has earned respect far beyond the shores of the British Isles, where it has successfully conserved all the breeds that have fallen within its remit: none have been lost since the Trust was formed. It has fulfilled its promise but its present stature and authority could hardly have been envisaged by the small body of far-sighted people who sparked off the concept of an organisation to conserve British rare breeds in the 1960s.

Five of that original group (Forwood, Stanley, Wheatley-Hubbard, Jewell and Rowlands) remain vice presidents of the Trust that they founded, and they let it be known that they are proud to be associated with it. They epitomise the firm thread of continuity that has counterbalanced the magnitude of the changes in the Trust's work and organisation since its inception — a thread strengthened by the loyalty and dedication over long periods by officers of the Trust and members of its Council. The Trust's debt to them all is incalculable. Without them, many breeds could have been lost forever.

Nor is the future being neglected. A great deal has been achieved but much remains to be done and will be in the hands of a new generation of leaders who have quietly accepted greater responsibilities in recent years. The pace of growth and development shows no sign of abating: the early 1990s have seen a renewed phase of expansionary policies, including a widening technical programme, valuable scientific research, a larger Show & Sale, the adoption of poultry conservation programmes and an interest in old traditional types of more popular breeds. There is also an increasing commitment to ensure that the Trust has sufficient political influence to protect rare breeds from potentially damaging legislation, both domestic and European, and to take advantage of the current emphasis on ecological issues and environmental considerations to which rare breeds can so richly contribute.

Environment; ecology; conservation; legislation; politics; science; technical projects; genetics — yes, all fine and important words. But a vital one is missing from that list, a word that describes more than any of them the driving force behind the work of all who care about rare breeds. The word is enthusiasm. It was enthusiasm and a sense of heritage, more than detached scientific interest or the practical demands of agriculture, that inspired the pioneers in the rare breeds movement; and the emotion of individuals is what truly drives a charity such as the Trust.

On a less insular note, there is as much passion for rare breeds among some North Americans as there is on this side of the Atlantic — and by no means all of it finds expression through the formality of an organisation. John Hughes, prolific script-writer of films such as *Home Alone*, lives in a suburb of Chicago but he also has a rare breeds farm in the country, which gives him far more pleasure than his lucrative film career. In a recent interview, he said: 'There are much more important things in life than movies. When I'm old and grey and battered I don't want to be still making them. I want to be doing what I should be doing, which is writing essays about wool or the preservation of the Devon milking cow.'

The Rare Breeds Survival Trust has now come of age. When it is old and grey it, too, will be doing what it should be doing. The breeds of the past have a future and the Trust will be there for them.

APPENDICES

MIDDLE WHITE GILT

APPENDIX I
RARE BREEDS ACCEPTANCE PROCEDURE

Section A: Genetic Basis

1. Has there been an accepted Stud/Herd/Flock Book for at least six generations?

2. Have other breeds contributed less than 20 per cent of the genetic make-up of the breed in the last six generations?

3. Are the parent breeds used in the formation of this breed no longer available?

4. Has it existed continuously for 75 years?

(If the answer to 4 is affirmative and if two or three of the other answers are affirmative, proceed to Section B.
If the breed does not qualify under this section, it is not accepted unless it possesses a distinct characteristic not found elsewhere.)

Section B: Numerical Basis

1. A breed will be included in the lists if there are fewer than the following number of breeding females in the breed:
Cattle 750; Horses 1,000; Sheep 1,500; Goats 500; Pigs 500.

2. Breeds with four or fewer distinct male lines will be included on the priority list. A watching brief will be kept on breeds with six or fewer distinct male lines. (A distinct male line is one which has no ancestors in common with another in the last four generations, i.e. up to and including great grandparents.)

Section C: Current Trends

1. Are the numbers of the breed decreasing significantly?

2. Is the breed found in fewer than four significant units which are more than 50 miles apart?

(Affirmative answers to these questions may give the breed a higher priority within the list, or may even permit a breed which does not qualify otherwise to be included.)

PRIORITY LIST CATEGORIES

Category 1: Critical
Category 2: Endangered
Category 3: Vulnerable
Category 4: At Risk
Category 5: Imported
Category 6: Feral

(Category 5: Sections A, B and C of the Rare Breeds Acceptance Procedure apply, but in addition the UK must be the main breeding centre for a breed that is seriously endangered, or extinct, in its country of origin.
Category 6: Feral populations exist in circumstances which limit the action that the RBST is able to take to ensure their conservation. Their acceptance onto the Priority List is determined by their numerical status and their historical and genetic significance.)

APPENDIX II
COUNCIL MEMBERS AND OFFICE HOLDERS

ALDERSON, G.L.H.
Working Party; Committee 1973; Technical Consultant 1973—90; Director of Show & Sale 1974—6; Editor *The Ark* 1976—82; Executive Director 1991—; Founder chairman RBI 1991—93

ALLAN, Dr R.J.P.
Council (co-opted) 1983—4, (ex officio) 1986—88

ANN, M.D.M.
Working Party; Committee 1973—75; Council (member) 1975

BAILEY, R.F.
Council (member) 1984—87, 1989—92

BALLANCE, Major J.C.R.
Council (member) 1990—93

BARBER of Tewkesbury, Lord
Council (co-opted) 1987—91; President 1991—

BLACK, J.A.C.
Council (co-opted) 1989—91, (member) 1991—

BOWMAN, Dr J.C., CBE
Working Party; Committee 1973—75; Council (member) 1975—77, (co-opted) 1977—84

BRADSHAW, J.E.
Council (member) 1990

BRIGGS, J.K.
Council (member) 1977—

BROOKS, B.R.
Council (member) 1987—93

CASSIDY, Mrs P.V.
Editor *The Ark* 1983—

CATOR, J.
Committee 1973—75; Council (member) 1975—81

CLOKE, G.E.
Council (member) 1976— ; Chairman 1983—86; Executive Chairman 1985; Chairman-elect 1992—94; Director of Trading Company 1984—

CLOKE, J.B.
Council (member) 1993—

CLOSE, P.W.
Council (member) 1988—91

CLUTTON-BROCK, Dr J.
Council (member) 1979—92

COLE-MORGAN, J.A.
Committee 1973—75

COLEMAN, A.C.
Council (co-opted) 1983, (member) 1983—90

COOK, K.N.
Council (co-opted) 1993—

COOPER, Sir Richard, Bt
Committee 1973—75; Council (member) 1975— ; Treasurer 1975—77; Chairman 1977—79, 1990—92

CRANBROOK, Earl of
Working Party; President 1973—76

DADD, C.V.T., OBE, FRAgS
Working Party; Committee 1973—75; Council (member) 1975—86

DEVONSHIRE, Duchess of
President 1977—79, 1988—89

DICKINSON, Mrs S.A.
Council (co-opted) 1992, (member) 1993—

DYMOND, A.J.
Chief Executive 1986—90; Field Director 1991

ELLIOTT of Morpeth, Lord
President 1989—91

FORWOOD, Sir Dudley, Bt
Task Conference chairman 1971; Committee 1973—75; Treasurer 1973—75; President 1980—82

FRAZER, Dr J.F.D.
Working Party; Committee 1973—75; Council (member) 1975—79

HARPER-SMITH, Dr J.R.
Council (member) 1989—

HAWTIN, A.J.
Council (member) 1979— ; Director of Show & Sale 1982—93

HENSON, Ms E.L.
Council (member) 1989—

HENSON, J.L.
Working Party; Chairman of Committee 1973—75; Chairman 1975—77, 1986—88; Council (member) 1975—88

HINDSON, J.C.
Council (member) 1975—88

HUNT, P.F.
Secretary 1973—78

JACKSON, G.H., OBE
Council (co-opted) 1981—

JEWELL, Professor P.A.
Working Party; Committee 1973—75; Council (member) 1975—91

LONGRIGG, W.
Working Party; Committee 1973—75; Council (member) 1975—80

LYONS, A.L.
Council (member) 1992—

MACK, Mrs C.M.
Council (member) 1987—

MANCHESTER, A.J.
Committee 1973—75; Council (co-opted) 1975—82

MARLER, C.J.S.
Council (member) 1976—89

MULHOLLAND, J.R.
Council (co-opted) 1985—

OTTER, R.G.
Council (member) 1975—86

PETCH, Mrs B.A.
Council (member) 1981—

PITMAN, Capt C.
Committee 1973—75

RAGG, Mrs R.M.
Director of Trading Company 1982—

149

REEVES, D.W.
Council (member) 1984— ; Chairman 1992—94

ROBERTS, C.J.
Council (member) 1992—

ROBINSON, A.J.
Council (member) 1987—90

ROSENBERG, M.M., CBE
Editor *The Ark* 1974—75; Council (member) 1975—86; Director of Show & Sale 1976—81; Chairman 1979—81; Hon Director 1982—85; Director of Trading Company 1982—88; President 1987—88

ROWLANDS, Dr I.W.
Working Party; Committee 1973—75; Council (member) 1975—80

RYDER, Dr M.L.
Council (member) 1975—6

SHEPPY, A.J.
Committee 1974—75; Council (member) 1975—79, 1991—

SNELL, P.G.
Council (member) 1986—89

STANLEY, J.W.
Council (member) 1993—

STANLEY, W.F., OBE, FRAgS
Chairman of Working Party 1969—72

STEANE, D.E.
Council (co-opted) 1981—89

STUBBS, D.H.
Council (member) 1977—87

TAYLOR, J.A.
Council (member) 1975—77

TERRY, R.
Secretary/Administration Director 1988— ; Director of Trading Company 1993—

TINKER, J.
Council (co-opted) 1989—

TITLEY, P.E.
Council (ex officio) 1986—88, (member) 1988—

VERNON, D.S.
Council (member) 1975— ; Treasurer 1978— ; Director of Trading Company 1982—92; Chairman 1988—90

WALLACE, L.A.
Council (member) 1980—83

WALTERS, Dr J.R.
Council (co-opted) 1991— ; RBI Advisory Council 1992—

WATSON, J.M.
Council (member) 1980—82

WELLINGTON, Duke of, KG, MVO, OBE, MC, DL
President 1983—87

WHEATLEY-HUBBARD, Mrs E.R., OBE, FRAgS
Working Party vice chairman; Committee 1973—75; Council (member) 1975—84; Acting Chairman 1976; Chairman 1981—83

WOOD-ROBERTS, J.H.
Secretary 1983—88

Presidents

1973—76 The Earl of Cranbrook
1977—79 The Duchess of Devonshire
1980—82 Sir Dudley Forwood, Bt
1983—87 The Duke of Wellington
1987—88 M.M. Rosenberg
1988—89 The Duchess of Devonshire
1989—91 Lord Elliott of Morpeth
1991—   Lord Barber of Tewkesbury

Chairmen

1973—77 J.L. Henson
1977—79 R.P. Cooper
1979—81 M.M. Rosenberg
1981—83 Mrs E.R. Wheatley-Hubbard
1983—86 G.E. Cloke
1986—88 J.L. Henson
1988—90 D.S. Vernon
1990—92 Sir Richard Cooper, Bt
1992—94 D.W. Reeves

LONG SERVICE

1. Members of Council

15 years or more:	10 years or more:
* J.K. Briggs	Dr J. Clutton-Brock
* G.E. Cloke	C.V.T. Dadd, OBE
* Sir Richard Cooper, Bt	* A.J. Hawtin
Prof P.A. Jewell	J.L. Henson
* D.S. Vernon	J.C. Hindson
	C.J.S. Marler
	R.G. Otter
	* Mrs B.A. Petch
	M.M. Rosenberg, CBE

(* = existing member of Council)

2. Officers

G.L.H. Alderson: Technical Consultant 17 years
D.S. Vernon: Treasurer 16 years
J.K. Briggs: Chairman Linga Holm subcommittee 16 years
G.E. Cloke: Chairman Breed Liaison committee 14 years
Mrs P.V. Cassidy: Editor *The Ark* 11 years
A.J. Hawtin: Chairman Show & Sale subcommittee 11 years

CURRENT MEMBERS OF STAFF

Executive Director: Lawrence Alderson
PA to Executive Director: Kay Burgess (with special responsibilities for co-ordinating Show & Sale and Show Demonstration programme)

Administration

Administration Director/Company Secretary: Robert Terry
PA to Administration Director: Val Nicholson (with special responsibilities for cattle and pig Semen Banks)
Membership: Linda Ridgeway, Valerie Dash (also reception)
Accounts: Pat Butler, Roy Brooks

Technical Department

Field Officer: Peter King
Livestock registration (CFB, BPA): Judy Milburn, Diane Parkinson

Publicity

Editor and Publicity Officer: Mrs Pat Cassidy
PA to the Editor: Linn Steele
Merchandising Officer: Mrs Rosalind Ragg

APPENDIX III
GRANT-GIVING TRUSTS AND SPONSORS

The following have supported the Trust:

Esme Fairbairn Charitable Trust
Cecil Pilkington Charitable Trust
Sun Alliance Group
The Rare Breeds Survival Trust American Foundation
Pamela Sheridan Charitable Trust
Ernest Cook Trust
The Dulverton Trust
Dumbreck Charity
Christopher H.R. Reeves Charitable Trust
Daihatsu (UK) Ltd
NFU Mutual
TMAgricom
Volvo Concessionaires
Butlers Fuels
Upjohn Ltd
A.W. Craiggy Charitable Trust
Whitley Animal Protection Trust
Glemco Trust
The Wilfred and Constance Cave Foundation
The Sir William Lyons Charitable Trust
Country Living Magazine
Lloyds Bank
National Westminster Bank
Pitman Moore
Jameson Whiskey
Hudson & Middleton
Deosan
Landini
Youngs Animal Health
M & M Timber
NERC
William A. Cadbury Charitable Trust
The Joicey Trust
Hadrian Trust
Loke Wan Tho Memorial Trust
Godfrey Mitchell Charitable Trust
Yorkshire Agricultural Society
The Esme Mitchell Trust
The Jane Hodge Foundation
Boots Charitable Trust
R.D. Turner Charitable Trust
The A.M. Fenton Trust
P.F. Charitable Trust
Sir Andrew Cornwath Charitable Trust
Worshipful Company of Grocers
The Rothley Trust
The John and Daphne Ward Charitable Trust
The Nichols Charitable Trust
Wellington Charitable Settlement
Clermont Charitable Co
Nigel Vincent Charitable Trust
Broughton Trust
Robbins Hill Trust

and many private individuals

APPENDIX IV
CORPORATE MEMBERS

AB Hoses & Fittings Ltd
Aldenham Country Park
Avoncroft Cattle Breeders
BP Nutrition Ltd
Belfast Zoo
J. Bibby Agriculture Ltd
BOCM Pauls Ltd
Bramley & Wellesley Ltd
British Waterfowl Association
Buccleuch Countryside Service
Butler Fuels
Canterbury Oast Trust
Cape plc
Carlton Communications plc
James Chapman (Butchers) Ltd
Cholderton Rare Breeds Frm
Ciba Geigy Agriculture
City of Sheffield Recreation Department (Graves Park)
Clifford Chance
Cotswold Farm Park
Alfred Cox (Surgical) Ltd
Crawley Borough Council (Tilgate Park)
Croxteth Hall and Country Park
Cyanamid of Great Britain Ltd
Dalgety plc
Danco plc
Davies Jones & Holland, MsRCVS
Deosan Ltd
Devon County Agricultural Association
Ditchingham Farms
Dixons Group plc
Domestic Fowl Trust
East of England Agricultural Society
East Hele Farm
FarmWorld
Farmers Weekly
Ferguson International Holdings plc
Folly Farm Waterfowl
Foundation for Water Research
GENUS
Gillhouse Herd of Pedigree Pigs
Glaxo Group Research Ltd
Hambros Bank Ltd
Harris Associates Ltd (formerly Exporc)
Hoddom and Kinmount Estates
Home Farm Temple Newsam (Leeds City Council)
Hoechst UK Ltd
Instituto de Zootecnia, Cordoba
Intervet UK Ltd
Janssen Animal Health
Kemira Fertilisers
Kent and Surrey Commons and Burnham Beeches
Ketchum Manufacturing Co
Kirtlington Stud
Isa Lloyd Animal Systems
Lloyds Bank plc
London Borough of Redbridge (Hainault Forest Country Park)
S. McGettigan & A. McGettigan, MsRCVS
National Trust (Wimpole Home Farm)
New Forest Agricultural Society

Northern Bank Ltd
Osmonds Ltd
Parke-Davis Veterinary
Pig Genetics
Pitman-Moore Ltd
Mrs P. Quinn
Rare Farm Animals of Hollanden
Reddaways Livestock Transport
Royal Agricultural Society of England
Royal Bath and West of England Society
Royal Cornwall Agricultural Association
Royal Smithfield Club
Royal Ulster Agricultural Society
Rumenco Ltd
Safeway plc
Sherwood Forest Farm Park
Shetland Sheep Breeders' Group
South of England Agricultural Society
Staffordshire County Council (Shugborough Estate)
Sun Alliance Insurance Group
Surrey County Agricultural Society
Themed Leisure Services Ltd
The Orkney Pig Project
The Scottish Gourmet
The Showman's Directory
The Wilfred & Constance Cave Foundation
John D. Thornborrow, FRICS
Three Counties Agricultural Society
TM Agricom Ltd
Toddington Manor Farms
Trustees of the Bedford Estate
Turnbull & Jackson, MsRCVS
Veterinary Life Assurance Services Ltd (VETLAS)
Volac Ltd
Votex Hereford Ltd
Weetabix Ltd
Whitbread Hop Farm
Yorkshire Agricultural Society
Young's Animal Health Ltd

GLOSSARY OF ACRONYMS

ABRO	Animal Breeding Research Organisation
AC	Advisory Committee (RBST)
ADAS	Agricultural Development and Advisory Service
AI	artificial insemination
ALHFAM	Association of Living Historical Farms and Museums
AMBC	American Minor Breeds Conservancy
ARC	Agricultural Research Council
BPA	British Pig Association (formerly NPBA)
BPC	Breeding Policy Committee (ZSL)
BWA	British Waterfowl Association
CAS	Centre for Agricultural Strategy
CFB	Combined Flock Book
CLA	Country Landowners Association
DAFS	Department of Agriculture for Scotland
DANI	Department of Agriculture for Northern Ireland
EAAP	European Association for Animal Production
EBL	enzootic bovine leukaemia
EC	European Community
f&m	foot and mouth disease
FAO	Food and Agriculture Organisation (United Nations)
FPS	Fauna Preservation Society
FWAG	Farming and Wildlife Advisory Groups
GBSC	Gene Bank subcommittee (of the ZSL's BPC)
GOS	Gloucester Old Spots (pigs)
HFRO	Hill Farming Research Organisation
MAFF	Ministry of Agriculture, Fisheries and Food
MERL	Museum of English Rural Life (now RHC)
MLC	Meat and Livestock Commission
MMB	Milk Marketing Board
NAAS	National Agricultural Advisory Service (now ADAS)
NAC	National Agricultural Centre (RASE)
NCBA	National Cattle Breeders Association
NERC	Natural Environment Research Council
NFU	National Farmers Union
NPBA	National Pig Breeders Association (now BPA)
OSB	Oxford Sandy and Black (pig)
RASE	Royal Agricultural Society of England
RBI	Rare Breeds International
RBST	Rare Breeds Survival Trust (originally Task)
RBWP	Rare Breeds Working Party
RHC	Rural History Centre (formerly MERL)
RSA	Royal Society of Arts
RSPB	Royal Society for the Protection of Birds
RSPCA	Royal Society for the Prevention of Cruelty to Animals
RZSS	Royal Zoological Society of Scotland
SAP	Scientific Advisory Panel (RBST)
SSC	Standing subcommittee (RBST)
TAG	Technical Advisory Group (RBST)
UFAW	Universities Federation for Animal Welfare
ZSL	Zoological Society of London

BIBLIOGRAPHY

Alderson, G.L.H.
  The Observer's Book of Farm Animals
    (Frederick Warne, 1976)
  Rare Breeds (Shire Publications, 1984)
  The Chance to Survive (revised edn, 1989),
    A.H. Jolly (Editorial) Ltd.
  Genetic Conservation of Domestic Livestock (ed.)
    (CAB International, 1990)

Alderson, G.L.H. and Bodó, I. (eds)
  Genetic Conservation of Domestic Livestock,
    Vol. II (CAB International, 1992)

American Minor Breeds Conservancy
  American Minor Breeds Notebook (1989)

Bowman, J.C., and Aindow, C.A.
  *Genetic Conservation and the Less Common
    Breeds of British Cattle, Pigs and Sheep*
    (University of Reading, Department of
    Agriculture and Horticulture, Study No.
    13, 1973)

Clutton-Brock, J.
  Domesticated Animals (Heinemann/BMNH,
    1981)

Curran, P.L.
  *Kerry and Dexter Cattle and Other Ancient
    Irish Breeds: A History* (Royal Dublin
    Society, 1990)

Elwes, Henry John
  *A Guide to the Primitive Breeds of Sheep and
    their Crosses* (R. and R. Clark, Edinburgh,
    1913)

FAO
  *Pilot Study on the Conservation of Animal
    Genetic Resources* (Rome, 1975)

Goddard, Nicholas
  *Harvests of Change: The Royal Agricultural
    Society of England, 1838-1988* (Quiller
    Press, 1988)

Hall, S., and Clutton-Brock, J.
  Two hundred years of British farm livestock
    (British Museum (Natural History), 1989)

Hodges, John, Simon, D.L., Smith, Charles *et al.*
    (eds)
  Conservation of Animal Genetic Resources
    (EAAP Livestock Production Science,
    11, 1984)

Jewell, P.A., et al. (eds)
  *Island Survivors: The Ecology of the Soay Sheep
    of St Kilda* (Athlone Press, 1974)

Mason, I.L.
  *A World Dictionary of Livestock Breeds, Types
    and Varieties* (3rd rev. edn, CAB Interna-
    tional, 1988)

Porter, Valerie
  Practical Rare Breeds (Pelham, 1987)
  The Southdown Sheep (Weald & Downland
    Open Air Museum, 1991)
  Cattle: A Handbook to the Breeds of the World
    (Christopher Helm (Publishers) Ltd,
    1991)

  Pigs: A Handbook to the Breeds of the World
    (Helm Information, 1993)

Royle, Nicola J.
  'Polymorphisms in rare breeds of cattle' (PhD
    thesis, Reading, 1983)

Ryder, M.L.
  Sheep and Man (Duckworth, 1983)

Speed, J.G. and M.G.
  The Exmoor Pony (Countrywide Livestock,
    1977)

Stuart, Lord David
  An Illustrated History of Belted Cattle (Scottish
    Academic Press, Edinburgh, 1970)

Stout, Adam
  The Old Gloucester (Alan Sutton Publishing,
    1980)

Tribe, D.E. and E.M.
  'North Ronaldsay sheep' (*Scottish Agriculture*,
    29 (1949))

Trow-Smith, R.
  A History of British Livestock Husbandry (2 vols,
    Routledge & Kegan Paul, 1957 and 1959)

Wade-Martins, Peter
  The Manx Loghtan Story (Geerings of Ashford,
    1990)
  *Black Faces - A History of East Anglian Sheep
    Breeds* (Norfolk Museum Service/Geerings
    of Ashford)

Werner, A.R.
  *An Enquiry into the Origin of Piebald or Jacob
    Sheep* (reprint, Countrywide Livestock,
    1975)

Whitehead, G.K.
  *The Ancient White Cattle of Britain and their
    Descendants* (Faber & Faber, 1953)

Wiseman, J.A.
  A History of the British Pig (Duckworth, 1986)

INDEX

INDEX